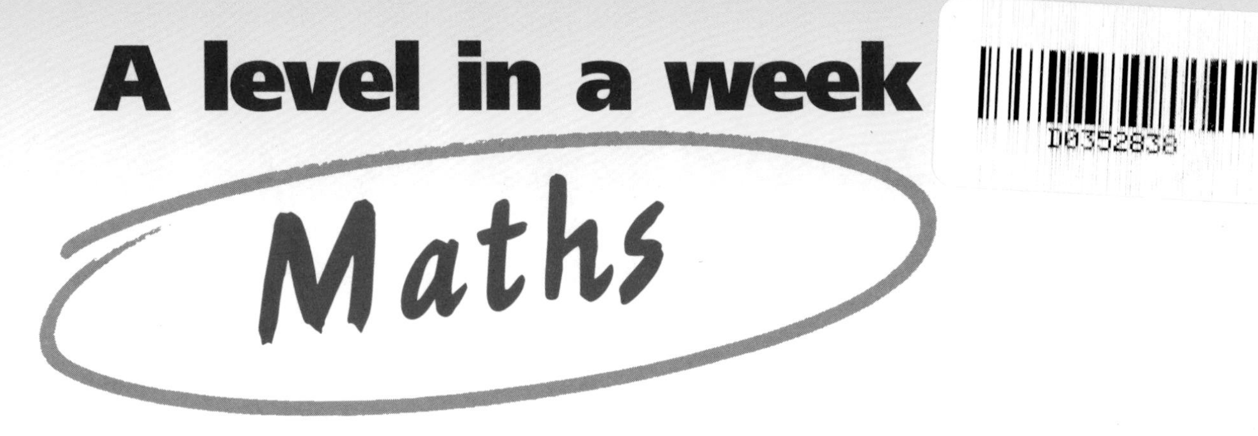


**Catherine Brown and Lee Cope**
**Series editor: Kevin Byrne, Abbey Tutorial College**

## *Where to find the information you need*

Letts Educational
Aldine Place
London W12 8AW
Tel: 0181 740 2266
Fax: 0181 743 8451
e-mail: mail@lettsed.co.uk
website: http://www.lettsed.co.uk



First published 1999



**British Library Cataloguing in Publication Data**
A CIP record for this book is available from the British Library.

ISBN 1 85758 9300

Editorial, design and production by Hart McLeod, Cambridge

Printed in Great Britain by Ashford Colour Press

Letts Educational is the trading name of BPP (Letts Educational) Ltd

# Algebra 1

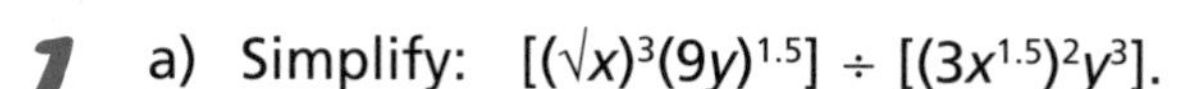


**1**
a) Simplify: $[(\sqrt{x})^3(9y)^{1.5}] \div [(3x^{1.5})^2y^3]$.
b) i) Express as powers of 3: $9^x$, $27^{1-4x}$.
ii) Hence solve the equation $3^{2-3x}9^x = 27^{1-4x}$.

**2** Express the following in terms of the simplest possible surds:

a) $\sqrt{343}$  b) $\sqrt{768}$  c) $\dfrac{\sqrt{3}}{2\sqrt{6}}$  d) $\dfrac{3}{1-\sqrt{7}}$.

**3**
a) Express as a single logarithm: $\log 2 + \log 6 - 2\log 3$.
b) Express in terms of simple logarithms: $\log\left(\dfrac{a^2b^3}{c}\right)$.
c) Make y the subject of the following:
i) $\ln y + 2\ln x = 2\ln 4$  ii) $2e^{y+1} = x + 1$.
d) Solve the equation: $8^x = 3$ giving your answer correct to two decimal places.

**4** Solve the following inequalities:
a) $3x - 5 \geq 4x - 8$  b) $4x^2 - 8x + 3 > 0$.

**5**
a) Solve the following equations:
i) $|4 - 2x| = 10$  ii) $|x - 3| = |2x + 1|$.
b) Sketch the graph of $y = |x + 4|$

## Answers

**1.** a) $3x^{-1.5}y^{-1.5}$ b) i) $9^x = 3^{2x}$; $27^{1-4x} = 3^{3-12x}$ ii) $x = \frac{1}{11}$

**2.** a) $7\sqrt{7}$ b) $\sqrt{768} = 16\sqrt{3}$ c) $\dfrac{\sqrt{2}}{4}$ d) $\dfrac{-(1+\sqrt{7})}{2}$

**3.** a) $\log(\frac{4}{3})$ b) $2\log a + 3\log b - \log c$
c) i) $y = \frac{16}{x^2}$ ii) $y = \ln[\frac{1}{2}(x + 1)] - 1$
d) $x = 0.53$

**4.** a) $x \leq 3$ b) $x < \frac{1}{2}$ or $x > \frac{3}{2}$

**5.** a) i) $x = -3$ or $x = 7$ ii) $x = \frac{2}{3}, -4$ b)

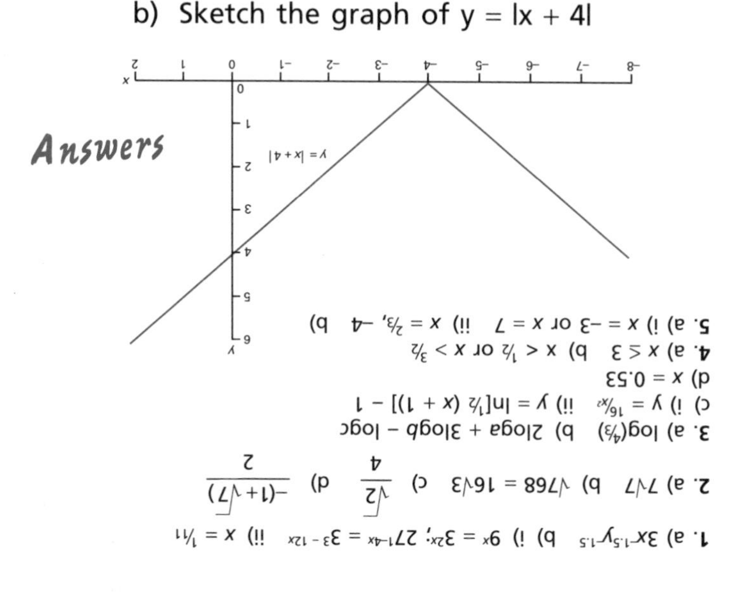


If you got them all right, skip to page 8

# Algebra 1

**1** You need to learn the **laws of powers**:

$y^m \times y^n = y^{m+n}$ $y^m \div y^n = y^{m-n}$ $(y^m)^n = y^{mn}$

In other words: if you are multiplying, add the powers; if you are dividing, subtract them; and if you have a power of a power, multiply.

In addition to the laws, you need to know that:

*You can't use these to deal with something like $2^4 \times 5^3$, because you must have the same base for each power*

- Anything to the power 0 is 1.
- Anything to the power 1 is the number itself – $5^1 = 5$.
- A negative power means 'one over' $x^{-3} = {}^1\!/\!_{x^3}$
- A fraction power means roots $x^{\frac{1}{2}} = \sqrt{x}$; $x^{\frac{1}{3}} = \sqrt[3]{x}$; $x^{\frac{3}{2}} = \sqrt{x^3}$ or $\left(\sqrt{x}\right)^3$.

You can manipulate powers only in certain ways. You can take out brackets when only × or ÷ are involved, but not with + or –:

$(2x)^3 = 2^3x^3 = 8x^3$; $(5/3)^4 = 5^4/3^4$ but $(x + a)^{0.5}$ is definitely **not** $x^{0.5} + a^{0.5}$.

The examples below illustrate two ways of using these facts about powers; powers are essential in many other areas too.

**Example 1**

a) Express i) $2^{4x-6}$ ii) $16^{5x-3}$ iii) $8^{2x-4}$ in the form $4^y$, where $y$ is to be determined in terms of $x$.

b) Hence solve the equation $\dfrac{2^{4x-6}}{4^{2-x}} = \dfrac{16^{5x-3}}{8^{2x-4}}$

**Solution**

a) i) Set $2^{4x-6} = 4^y$ (since we are told to). To solve this, we need both sides as powers of same number. Use $4 = 2^2$:

$$2^{4x-6} = 4^y \Rightarrow 2^{4x-6} = (2^2)^y = 2^{2y} \Rightarrow 4x - 6 = 2y \Rightarrow y = 2x - 3 \Rightarrow 2^{4x-6} = 4^{2x-3}$$

ii) $16^{5x-3} = 4^y \Rightarrow (4^2)^{(5x-3)} = 4^y \Rightarrow 4^{10x-6} = 4^y \Rightarrow y = 10x - 6 \Rightarrow 16^{5x-3} = 4^{10x-6}$

iii) $8^{2x-4} = 4^y \Rightarrow (2^3)^{2x-4} = (2^2)^y \Rightarrow 2^{6x-12} = 2^{2y} \Rightarrow 6x - 12 = 2y \Rightarrow y = 3x - 6$
$\Rightarrow 8^{2x-4} = 4^{3x-6}$

b) We must use what we've done: $\dfrac{2^{4x-6}}{4^{2-x}} = \dfrac{16^{5x-3}}{8^{2x-4}} \Rightarrow \dfrac{4^{2x-3}}{4^{2-x}} = \dfrac{4^{10x-6}}{4^{3x-6}}$

*You can't cancel the 4s at this stage*

Since we are dividing, we subtract powers:

$4^{2x-3-(2-x)} = 4^{10x-6-(3x-6)} \Rightarrow 4^{3x-5} = 4^{7x} \Rightarrow 3x - 5 = 7x \Rightarrow x = -5/4.$

**Example 2**

a) Given that $y = 5^x$, show that i) $125^x = y^3$ ii) $25^x = y^2$ iii) $5^{x+1} = 5y$.

b) Hence solve the equation $125^x - 6(25^x) + 5^{x+1} = 0$.

**Solution**

a) i) $125^x = (5^3)^x = 5^{3x}$ $y^3 = (5^x)^3 = 5^{3x}$ So $125^x = y^3$

ii) $25^x = (5^2)^x = 5^{2x}$ $y^2 = (5^x)^2 = 5^{2x}$ So $25^x = y^2$

iii) $5^{x+1} = 5^x \times 5^1$ $5y = 5 \times 5^x = 5^x \times 5^1$ So $5^{x+1} = 5y$.

b) Using what we've done:

$125^x - 6(25^x) + 5^{x+1} = 0 \Rightarrow y^3 - 6y^2 + 5y = 0 \Rightarrow y(y^2 - 6y + 5) = 0$
$\Rightarrow y(y-5)(y-1) = 0$

So $y = 0$, or $y = 5$, or $y = 1$. But $y = 5^x \Rightarrow 5^x = 0$, or $5^x = 5$, or $5^x = 1$.
$5^x = 0$ gives no solution, $5^x = 5$ gives $x = 1$ and $5^x = 1$ gives $x = 0$.

**2** A **surd** is something with a square root in it, like $\sqrt{2}$ or $\sqrt{5}$. You need to be able to manipulate surds in the following ways:

- **Simplifying**: when you have the square root of a number and you have to express it in terms of the simplest possible surds – often – $\sqrt{2}$, $\sqrt{3}$ or $\sqrt{5}$. Example 3 illustrates this:

**Example 3**

Express in terms of the simplest possible surds: a) $\sqrt{8}$ b) $\sqrt{243}$

**Solution**

a) We look for a square number that goes into 8: 4 does.

So write: $\sqrt{8} = \sqrt{(4 \times 2)} = \sqrt{4} \times \sqrt{2} = 2\sqrt{2}$

b) Find a square number again: 9 goes into 243.

So write: $\sqrt{243} = \sqrt{(9 \times 27)} = \sqrt{9} \times \sqrt{27} = 3\sqrt{27}$

But we have not finished, because we can find a square number that goes into 27: it's 9 again:

So $\sqrt{243} = 3\sqrt{27} = 3\sqrt{(9 \times 3)} = 3\sqrt{9} \times \sqrt{3} = 3 \times 3\sqrt{3} = 9\sqrt{3}$.

- **Rationalising the denominator**: getting surds off the bottom of a fraction. This is done by multiplying the top and bottom of the fraction by the same thing. There are two cases, according to what the denominator is like. These are illustrated in Example 4:

**Example 4**

Rationalise: a) $\frac{2}{3\sqrt{5}}$ b) $\frac{4-2\sqrt{6}}{2+\sqrt{3}}$.

**Solution**

a) Multiply top and bottom by $\sqrt{5}$: $\frac{2}{3\sqrt{5}} \times \frac{\sqrt{5}}{\sqrt{5}} = \frac{2\sqrt{5}}{3\sqrt{5}\sqrt{5}} = \frac{2\sqrt{5}}{3 \times 5} = \frac{2\sqrt{5}}{15}$

b) Multiply top and bottom by $2 - \sqrt{3}$:

$$\frac{(4-2\sqrt{6})}{(2+\sqrt{3})} \times \frac{(2-\sqrt{3})}{(2-\sqrt{3})} = \frac{(4-2\sqrt{6})(2-\sqrt{3})}{(2+\sqrt{3})(2-\sqrt{3})} = \frac{8-4\sqrt{3}-4\sqrt{6}+2\sqrt{6}\sqrt{3}}{4-2\sqrt{3}+2\sqrt{3}-\sqrt{3}\sqrt{3}}$$

$$= \frac{8-4\sqrt{3}-4\sqrt{6}+2\sqrt{18}}{4-3} = 8-4\sqrt{3}-4\sqrt{6}+2\sqrt{9}\sqrt{2} = 8-4\sqrt{3}-4\sqrt{6}+6\sqrt{2}.$$

**3** You need to know the laws of **logarithms**:

$\log(ab) = \log a + \log b$  $\log(a/b) = \log a - \log b$  $\log(a^n) = n\log a$

In other words: if you are multiplying the numbers, add the logs; if you are dividing, subtract them and you can 'bring down' powers to the front of the log.

You also need to know that:

- $\log 1 = 0$
- $\log_a a = 1$, whatever $a$ (the base of the logarithm) is
- You cannot have logs of negative numbers, or zero.

You can manipulate logs **only** according to these rules – you can do **nothing** with log(a + b), for example.

You will mainly deal with lnx, which is the 'opposite' (or more formally, the inverse function of) $e^x$. This means that $e^{\ln x} = \ln e^x = x$ so e and ln 'cancel'. You can use this to get rid of ln or e in an equation – but you have to be very careful how you do it! This is illustrated in Example 5:

*This cancelling only works when there is nothing in between the e and ln – you cannot 'cancel' $e^{-\ln x}$ or $\ln(2e^x)$ directly.*

**Example 5**

a) Given that $\ln y = 2\ln x + \ln 5$, express $y$ in terms of $x$.

b) Given that $e^{3y+4} - 5 = 2x$, express y in terms of $x$.

**Solution**

a) Before trying to get rid of the lns, we need to get both sides in the form ln(something).

We know $2\ln x = \ln x^2$ (laws of logs)
So $\ln y = \ln(x^2) + \ln 5 = \ln(5x^2)$ (laws of logs)

Now we can take exponentials: $e^{\ln y} = e^{\ln(5x^2)} \Rightarrow y = 5x^2$

b) We will need to take logs to get rid of the e, but first we must make sure the $e^{\text{something}}$ is on its own:

$e^{3y+4} - 5 = 2x \Rightarrow e^{3y+4} = 2x + 5$.

Taking logs: $\Rightarrow \ln(e^{3y+4}) = \ln(2x + 5)$

So $3y + 4 = \ln(2x + 5) \Rightarrow y = \dfrac{\ln(2x+5)-4}{3}$.

*Brackets are vital – it is NOT ln2x + ln5*

Logs can also be used to solve equations with powers in them, as in Example 6:

**Example 6**

Solve the following equation $3^x = 4^{\frac{1}{x}}$; $x > 0$ giving your answer correct to two decimal places.

**Solution**

To 'bring powers down', take logs of each side:

$$\ln(3^x) = \ln(4^{\frac{1}{x}}) \Rightarrow x\ln 3 = \frac{1}{x}\ln 4$$

Rearranging: $x^2 = {}^{\ln 4}\!/\!_{\ln 3} \Rightarrow x = 1.12$ (two d.p.)

**4** When dealing with **inequalities**, you must remember the following rules:

- You can **add** or **subtract anything** whatsoever to either side
- You can **multiply** or **divide** by **positive** numbers
- You can **multiply** or **divide** by **negative** numbers, **provided you change the direction of the sign**
- You **cannot multiply** or **divide** by **unknowns** such as $x$
- You **cannot square, square root, turn** them **upside down** – or anything else!

**Linear inequalities** are easy! You treat them just like equations (but remember about multiplying/dividing by negatives).

**Example 7**

Solve the inequality $2x - 5 \leq 3x - 4$.

**Solution**

Take $x$ terms to one side, numbers to the other:

$$2x - 5 \leq 3x - 4 \Rightarrow 2x - 3x \leq -4 + 5 \Rightarrow -x \leq 1$$

To get rid of the minus sign, divide by –1 and change direction of sign:

$$-x \leq 1 \Rightarrow x \geq -1.$$

To deal with **quadratic inequalities**, you must first find the roots and then draw a number line:

**Example 8**

Solve the following inequalities:

a) $x^2 - 3x - 4 > 0$ b) $2x^2 - x \leq 1$

*Be careful with < or ≤*
*Getting them wrong will cost marks*

**Solution**

a) Follow these steps:

Step 1 Find roots of the equation: $x^2 - 3x - 4 = 0 \Rightarrow (x - 4)(x + 1) = 0 \Rightarrow x = 4, -1$

Step 2 Mark them on a number line: –1 4

Step 3 Choose values of $x$ each side of the roots and between them; put into quadratic and see whether +ve or –ve:

$x = -2$: $(-2)^2 - 3(-2) - 4 = 6$ +ve
$x = 0$: $(0)^2 - 3(0) - 4 = -4$ –ve
$x = 5$: $(5)^2 - 3(5) - 4 = 6$ +ve

Step 4 Mark the appropriate regions on the number line +ve or –ve: +ve –1 –ve 4 +ve

Step 5 Write down the answer: Since we wanted $x^2 - 3x - 4 > 0$, we want $x < -1$ or $x > 4$.

b) Before we can start, we must first rearrange the equation to get 0 on one side: $2x^2 - x \leq 1 \Rightarrow 2x^2 - x - 1 \leq 0$.

Step 1 $2x^2 - x - 1 = 0 \Rightarrow (2x + 1)(x - 1) = 0 \Rightarrow x = -\frac{1}{2}, 1$

Step 2 $-\frac{1}{2}$ 1

Step 3 $x = -1$: $2(-1)^2 - (-1) - 1 = 2$: +ve $x = 0$: $2(0)^2 - (0) - 1 = -1$: –ve
$x = 2$: $2(2)^2 - (2) - 1 = 1$: +ve

Step 4 +ve $-\frac{1}{2}$ –ve 1 +ve

Step 5 Since we wanted $2x^2 - x - 1 \leq 0$, we want $-\frac{1}{2} \leq x \leq 1$.

**5** The **modulus** of $x$ (written $|x|$) gives just the positive value of $x$.
$|3| = 3$ and $|-4| = 4$.

To solve an equation of the form modulus |something| = constant, let what is in the bracket equal plus or minus the constant. To solve an equation with a modulus sign each side, square both sides.

**Example 9**

Solve the equations a) $|2x - 5| = 1$ b) $|x - 3| = |3x - 1|$.

**Solution**

a) $|2x - 5| = 1 \Rightarrow 2x - 5 = 1$ or $2x - 5 = -1 \Rightarrow x = 3$ or $x = 2$

b) $|x - 3| = |3x - 1| \Rightarrow (x - 3)^2 = (3x - 2)^2 \Rightarrow x^2 - 6x + 9 = 9x^2 - 12x + 4$
$\Rightarrow 8x^2 - 6x - 5 = 0 \Rightarrow (2x + 1)(4x - 5) = 0 \Rightarrow x = -\frac{1}{2}, \frac{5}{4}$.

To draw the graph of the modulus of something, you draw the normal graph, and then 'reflect' the negative parts as in the examples below.

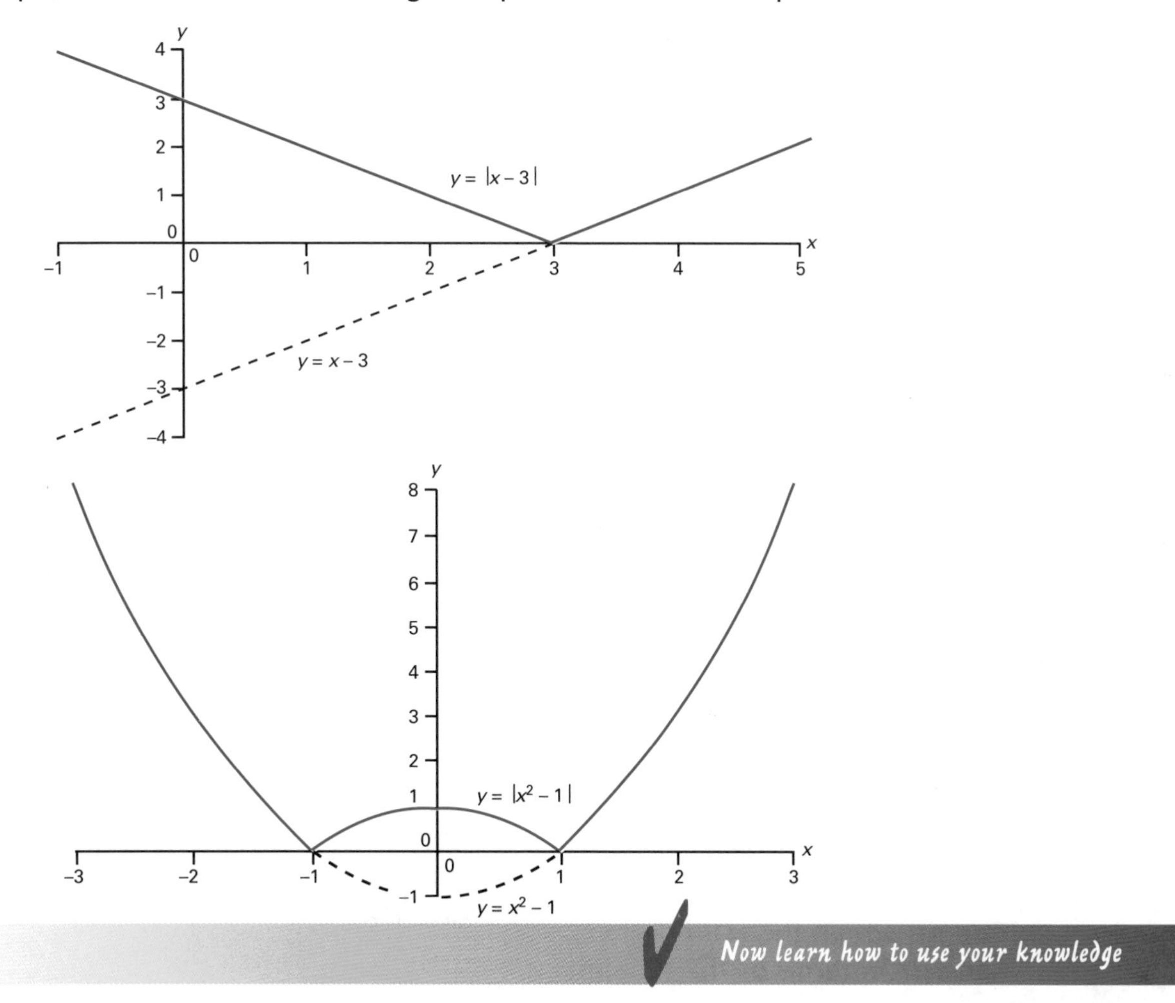

# Algebra 1

*Hints*

**1** a) Find the solutions to the quadratic equation $x^2 - 6x + 2 = 0$, giving your answers in terms of the simplest possible surds.

*Hint: Use the formula. Simplify the square root*

b) Hence obtain the exact solution to the inequality $x^2 - 6x + 2 \leq 0$.

*Hint: Even though it's surds, use normal method*

**2** A gardener is planning a rectangular lawn. He requires its width to be 4 m less than its length, its perimeter to be at most 36 m and its area to be at least 60 $m^2$. By forming and solving suitable inequalities, find the range of acceptable values for the length of the lawn.

*Hint: Call the length $x$ – and find width. Work out the perimeter and area. You need both inequalities to be true*

**3** A graph of $P$ against $Q$ produces a straight line with gradient –1 and intercept $2\ln 4$.

a) Write down an equation relating $P$ and $Q$.

b) Given that $P = \ln y$ and $Q = \ln x$, find y in terms of x, giving your answer in a form not involving logarithms.

*Hint: Substitute into your equation. Get each side in the form ln(something)*

**4** a) Sketch, on the same diagram, the graphs of $y = |2x - 1|$ and $y = 4x - 8$.

b) Hence, or otherwise, obtain the solution to the equation $|2x - 1| = 4x - 8$.

*Hint: Try squaring. How many solutions are there? Look at your graph*

c) Hence, or otherwise, obtain the solution to the inequality $|2x - 1| > 4x - 8$.

*Hint: Where is $2x - 1$ above $4x - 8$?*

**5** $f(x) \equiv 2x^2 + 20x - 15$.

a) Express f(x) in the form $A(x + B)^2 + C$, where A, B and C are constants to be determined.

*Hint: Expand out the brackets and equate coefficients*

b) Hence obtain the solution to the equation $f(x) = 31$, giving your answer in terms of the simplest possible surds.

**6** By using the substitution $y = 2^x$, obtain the solution to the equation $2^{x+1} + 2^{-x} = 3$.

**7** Obtain the solution to the equation $2e^{2x} - 11e^x + 15 = 0$, leaving your answer in terms of natural logarithms.

**Hints**

$2^{x+1} = 2^x \times 2^1$; $2^{-x} = \frac{1}{2^x}$

Let $y = e^x$
$e^{2x} = (e^x)^2$

Answers on page 85

# Algebra 2

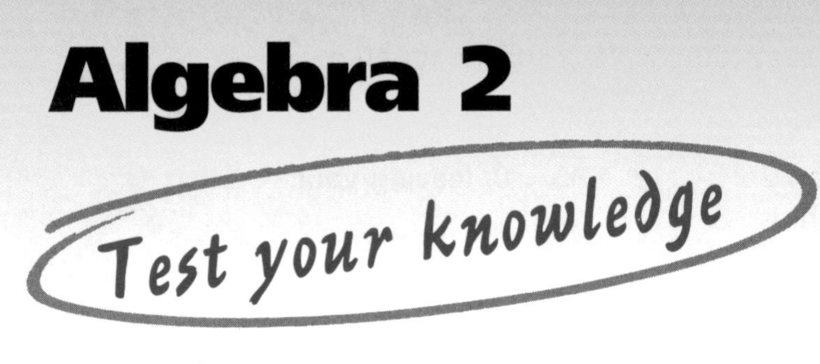

**1** a) Expand $(2 + 3x)^6$ up to the term in $x^3$.

b) Hence find an approximation to $2.03^6$.

**2** a) i) Find the remainder when $2x^3 - 2x^2 + 7x - 5$ is divided by $(x + 2)$.

ii) The remainder when $x^4 - 4x^3 + Ax - 5$ is divided by $(x - 1)$ is 1. Find the value of the constant A.

b) i) Show that $(2x - 1)$ is a factor of $8x^3 - 12x^2 + 6x - 1$.

ii) Factorise fully $x^3 - 4x^2 - 25x + 28$.

**3** Express in partial fractions:

a) $\dfrac{5x+37}{(x-1)(x+6)}$

b) $\dfrac{3x+14}{(x^2+6)(x-2)}$

c) $\dfrac{5x^2+21x+13}{(x+3)^2(x-2)}$.

## Answers

**1** a) $64 + 576x + 2160x^2 + 4320x^3$ b) $69.98032$

**2** a) i) $-43$ ii) $A = 9$

b) i) $8(\tfrac{1}{2})^3 - 12(\tfrac{1}{2})^2 + 6(\tfrac{1}{2}) - 1 = 1 - 3 + 3 - 1 = 0$ so $(2x - 1)$ is a factor

ii) $(x - 1)(x - 7)(x + 4)$

**3** a) $\dfrac{6}{(x-1)} + \dfrac{-1}{(x+6)}$ b) $\dfrac{(-2x-1)}{(x^2+6)} + \dfrac{2}{(x-2)}$ c) $\dfrac{1}{(x+3)^2} + \dfrac{2}{(x+3)} + \dfrac{3}{(x-2)}$

**1** The **Binomial Theorem** tells you how to expand a bracket of the form $(a + b)^n$ (a and b can be anything – numbers, letters, mixtures... ). It says (this should be in your formula book):

$$(a+b)^n \equiv a^n + \binom{n}{1}a^{n-1}b + \binom{n}{2}a^{n-2}b^2 + \ldots + \binom{n}{r}a^{n-r}b^r + \ldots + b^n$$

where $\binom{n}{r}$ means ${}^nC_r = \dfrac{n!}{r!(n-r)!}$

*There should be an ${}^nC_r$ button on your calculator – use it by putting n, then ${}^nC_r$, then r*

Examples 1, 2 and 3 illustrate the normal ways to use it.

**Example 1**

Find the expansion of $(2 - 3x)^4$.

**Solution**

We have $a = 2, \quad b = -3x, \quad n = 4.$

$(2 - 3x)^4 \equiv 2^4 + {}^4C_1\, 2^3(-3x) + {}^4C_2\, 2^2(-3x)^2 + {}^4C_3\, 2^1(-3x)^3 + {}^4C_4\, 2^0(-3x)^4$

*A good check is that the powers add up to n each time – so here, the powers add to 4*

You **must** put the brackets in! If you **don't** you are almost certain to get it **wrong**!

So $(2 - 3x)^4 \equiv 16 + (4)(8)(-3x) + (6)(4)(9x^2) + (4)(2)(-27x^3) + (1)(1)(81x^4)$

$\equiv 16 - 96x + 216x^2 - 216x^3 + 81x^4.$

**Example 2**

a) Find the first four terms in the expansion of $(4 + 2x)^{10}$.

b) Hence obtain an approximation to $4.02^{10}$.

**Solution**

a) $(4 + 2x)^{10} \equiv 4^{10} + {}^{10}C_1\, 4^9(2x) + {}^{10}C_2\, 4^8(2x)^2 + {}^{10}C_3\, 4^7(2x)^3 + \ldots$

So $(4 + 2x)^{10}$

$\approx 1048576 + (10)(262144)(2x) + (45)(65536)(4x^2) + (120)(16384)(8x^3)$

$\approx 1048576 + 5242880x + 11796480x^2 + 15728640x^3$

b) We must use first part. Let $(4 + 2x)^{10} = 4.02^{10}$
so $4 + 2x = 4.02$
so $2x = 0.02$
so $x = 0.01$

So put $x = 0.01$ in our expansion:

$$(4.02)^{10} \approx 1048576 + 5242880(0.01) + 11796480(0.01)^2 + 15728640(0.01)^3$$
$$\approx 1102200.17664$$

**Example 3**

a) Expand $(1 + 2x)^5$.

b) Write down the expansion of $(1 - 2x)^5$.

c) Hence find $(1 + 2\sqrt{2})^5 - (1 - 2\sqrt{2})^5$, giving your answer in surd form.

**Solution**

a) $(1 + 2x)^5$

$$\equiv 1^5 + {}^5C_1\ 1^4(2x) + {}^5C_2\ 1^3(2x)^2 + {}^5C_3\ 1^2(2x)^3 + {}^5C_4\ 1^1(2x)^4 + {}^5C_5\ 1^0(2x)^5$$
$$\equiv 1 + (5)(1)(2x) + (10)(1)(4x^2) + (10)(1)(8x^3) + (5)(1)(16x^4) + (1)(1)(32x^5)$$
$$\equiv 1 + 10x + 40x^2 + 80x^3 + 80x^4 + 32x^5.$$

b) We could do this again from scratch – but it is easier to just replace '$x$' with '$-x$' in what we've already worked out. Take care with brackets: $(-x)^2 = x^2$.

So $(1 - 2x)^5 \equiv 1 + 10(-x) + 40(-x)^2 + 80(-x)^3 + 80(-x)^4 + 32(-x)^5$
$\equiv 1 - 10x + 40x^2 - 80x^3 + 80x^4 - 32x^5.$

c) We must have $(1 + 2\sqrt{2})^5 = (1 + 2x)^5$, giving $x = \sqrt{2}$.

To find $(1 + 2\sqrt{2})^5 - (1 - 2\sqrt{2})^5$, we want:

$(1 + 2x)^5 - (1 - 2x)^5$
$\equiv (1 + 10x + 40x^2 + 80x^3 + 80x^4 + 32x^5) -$
$(1 - 10x + 40x^2 - 80x^3 + 80x^4 - 32x^5)$
$\equiv 20x + 160x^3 + 64x^5$

So $(1 + 2\sqrt{2})^5 - (1 - 2\sqrt{2})^5 \equiv 20(\sqrt{2}) + 160(\sqrt{2})^3 + 64(\sqrt{2})^5$

To find $(\sqrt{2})^3$ and $(\sqrt{2})^5$ exactly, use the laws of powers:
$(\sqrt{2})^3 \equiv (\sqrt{2})^2(\sqrt{2})^1 = 2\sqrt{2}$ and $(\sqrt{2})^5 \equiv (\sqrt{2})^4(\sqrt{2})^1 = 4\sqrt{2}$

So $(1 + 2\sqrt{2})^5 - (1 - 2\sqrt{2})^5 = 20(\sqrt{2}) + 160(2\sqrt{2}) + 64(4\sqrt{2})$
$= 20\sqrt{2} + 320\sqrt{2} + 256\sqrt{2} = 596\sqrt{2}.$

**2** The **Remainder Theorem** is about what remainder you get when you divide a polynomial (i.e. something with $x^3$, etc.) by a linear factor – something like $(x - 3)$ or $(2x - 1)$. It says:

**When you divide a polynomial $f(x)$ by $(bx - a)$, the remainder is $f(a/b)$.**

So if you were dividing by $(x - 3)$, you'd put $x = 3$; if by $(x + 1)$, you'd put $x = -1$ and if by $(2x - 1)$, you'd put $x = \frac{1}{2}$.

Examples 4 and 5 show how this is commonly used.

**Example 4**

a) Find the remainder when $x^4 - 3x^3 + 2x^2 - 5x + 6$ is divided by $(x + 2)$.

b) The remainder when $2x^3 - 3x^2 + 8x - A$ is divided by $(2x + 1)$ is $-3$. Find $A$.

**Solution**

a) To find the remainder, we put in $x = -2$:
$(-2)^4 - 3(-2)^3 + 2(-2)^2 - 5(-2) + 6 = 64$.

b) We put $x = -\frac{1}{2}$, and the answer must come to $-3$:
$2(-\frac{1}{2})^3 - 3(-\frac{1}{2})^2 + 8(-\frac{1}{2}) - A = -3 \Rightarrow -\frac{1}{4} - \frac{3}{4} - 4 - A = -3 \Rightarrow A = -2$.

**Example 5**

$f(x) \equiv x^3 + Ax^2 + Bx - 5$.

When $f(x)$ is divided by $(x - 1)$, the remainder is $-6$. When $f(x)$ is divided by $(x + 1)$, the remainder is 4. Find $A$ and $B$.

**Solution**

Put $x = 1$; answer comes to $-6$:

$(1)^3 + A(1)^2 + B(1) - 5 = -6 \Rightarrow A + B - 4 = -6 \Rightarrow A + B = -2$ (1)

Put $x = -1$; answer comes to 4:

$(-1)^3 + A(-1)^2 + B(-1) - 5 = 4 \Rightarrow A - B - 6 = 4 \Rightarrow A - B = 10$ (2)

So we have simultaneous equations. Adding (1) and (2):
$2A = 8 \Rightarrow A = 4$. Substituting back $\Rightarrow B = -6$.

The **Factor Theorem** is a special case of the remainder theorem. It says:

**$(x - a)$ a factor of $f(x) \Leftrightarrow f(a) = 0$.**

*Whenever you need to factorise a cubic, you must use this method*

We use this to factorise cubic equations.

- If you are asked to **show** something is a factor, you just substitute in the appropriate number and show the answer is equal to zero.

- If you have to **find** a factor, you guess ± numbers that go into the constant term and then put them in to see if you get zero.

Once you've got your first factor, you stop guessing and work out the rest.

**Example 6**

a) Show $(x + 1)$ is a factor of $x^3 + 3x^2 - 13x - 15$, and hence factorise this expression fully.

b) Factorise fully: $x^3 - 2x^2 + x - 2$.

*If your calculator factorises cubics, you must still show your working to get the marks.*

**Solution**

a) To show $(x + 1)$ is a factor, put in $x = -1$:
$(-1)^3 + 3(-1)^2 - 13(-1) - 15 = 0$ so $(x + 1)$ is a factor

Now set $x^3 + 3x^2 - 13x - 15 \equiv (x + 1)$(quadratic).
We need to find the terms of the quadratic.

First term must be $x^2$ (since $x \times x^2$ is needed to get $x^3$). Last term must be $-15$ (since $1 \times -15$ is needed to get $-15$).

Now we must find the middle term, which we call $Ax$, where $A$ is the number we want to find.

So $x^3 + 3x^2 - 13x - 15 \equiv (x + 1)(x^2 + Ax - 15)$.

We now look at the $x^2$ terms on both sides:

On the left, we have $3x^2$

On the right, we get $x^2$ terms from $1 \times x^2$ and $x \times Ax$

So $3x^2 \equiv x^2 + Ax^2 \Rightarrow 3 = 1 + A$ So $A = 2$.

So $x^3 + 3x^2 - 13x - 15 \equiv (x + 1)(x^2 + 2x - 15) \equiv (x + 1)(x + 5)(x - 3)$ (factorising the quadratic)

b) First guess a factor: we will guess $\pm 1$, $\pm 2$

$x = 1$: $(1)^3 - 2(1)^2 + (1) - 2 = -2$ so $(x - 1)$ not a factor
$x = -1$: $(-1)^3 - 2(-1)^2 + (-1) - 2 = -6$ so $(x + 1)$ not a factor
$x = 2$: $(2)^3 - 2(2)^2 + (2) - 2 = 0$ so $(x - 2)$ is a factor

So set $x^3 - 2x^2 + x - 2 \equiv (x - 2)(x^2 + Ax + 1)$
Look at $x^2$: $-2x^2 \equiv -2x^2 + Ax^2$ So $A = 0$
So $x^3 - 2x^2 + x - 2 \equiv (x - 2)(x^2 + 1)$. This cannot be factorised further, since $x^2 + 1$ has no factors.

**3** **Partial fractions** involve splitting up something like $\frac{3}{(x+1)(x-2)}$ into $\frac{1}{(x-2)} - \frac{1}{(x+1)}$.

There are three different types that you must know; the types are split according to what the denominator (bottom) of the fraction is like.

- Type I: **linear factors**: $\frac{3}{(x+1)(x-2)}$ Use $\frac{A}{(x-2)} + \frac{B}{(x+1)}$
- Type II: **squared inside bracket**: $\frac{3}{(x^2+2)(x-3)}$ Use $\frac{(Ax+B)}{(x^2+2)} + \frac{C}{(x-3)}$
- Type III: **squared outside bracket**: $\frac{3}{(x-1)^2(x+4)}$ Use $\frac{A}{(x-1)^2} + \frac{B}{(x-1)} + \frac{C}{(x+4)}$.

*In type II, the* ***(Ax + B)*** *must be over the denominator with the* $x^2$ *in it.*

In each case, you put the partial fractions over the denominator you started with (this is important), and substitute in values of $x$ to find the constants $A$, $B$ and $C$. Example 7 illustrates the method.

**Example 7**

Express as partial fractions:

a) $\frac{x-5}{(x-2)(x-3)}$ b) $\frac{2x^2+x+11}{(x^2+3)(x+1)}$ c) $\frac{x^2-9x+17}{(x-2)^2(x+1)}$.

**Solution**

a) $\frac{x-5}{(x-2)(x-3)} \equiv \frac{A}{(x-2)} + \frac{B}{(x-3)} \equiv \frac{A(x-3)+B(x-2)}{(x-2)(x-3)}$ So $x-5 \equiv A(x-3)+B(x-2)$

Now we choose values of $x$ to substitute in. Wherever possible, we try to make one of the brackets zero!

$x - 5 \equiv A(x - 3) + B(x - 2)$.

$x = 3$ $3 - 5 = A(3 - 3) + B(3 - 2)$. $\Rightarrow -2 = A(0) + B$ $\Rightarrow B = -2$

$x = 2$ $2 - 5 = A(2 - 3) + B(2 - 2)$. $\Rightarrow -3 = A(-1) + B(0)$ $\Rightarrow A = 3$.

b) $\frac{2x^2+x+11}{(x^2+3)(x+1)} \equiv \frac{(Ax+B)}{(x^2+3)} + \frac{C}{(x+1)} \equiv \frac{(Ax+B)(x+1)+C(x^2+3)}{(x^2+3)(x+1)}$

*Make sure you put the brackets around the* ***(Ax + B)*** *– or you'll make a mistake!*

So $2x^2 + x + 11 \equiv (Ax + B)(x + 1) + C(x^2 + 3)$.

$x = -1$ $2(-1)^2 + (-1) + 11 = (A(-1) + B)(-1 + 1) + C((-1)^2 + 3)$

$\Rightarrow 12 = (-A + B)(0) + C(4) \Rightarrow 12 = 4C \Rightarrow C = 3$

We cannot make the other bracket zero, so choose other easy values of $x$. **Always start with $x = 0$**:

$x = 0$: $2(0)^2 + (0) + 11 = (A(0) + B)(0 + 1) + C((0)^2 + 3)$
$\Rightarrow 11 = B(1) + C(3)$. But $C = 3 \Rightarrow 11 = B + 9 \Rightarrow B = 2$

$x = 1$ $2(1)^2 + (1) + 11 = (A(1) + B)(1 + 1) + C((1)^2 + 3)$
$\Rightarrow 14 = (A + B)(2) + C(4)$.
But $C = 3$, $B = 2 \Rightarrow 14 = 2A + 4 + 12 \Rightarrow A = -1$.

c) $$\frac{x^2 - 9x + 17}{(x-2)^2(x+1)} \equiv \frac{A}{(x-2)^2} + \frac{B}{(x-2)} + \frac{C}{(x+1)}.$$

With this type we must take **special** care putting them over a common denominator. The denominator we use is $(x - 2)^2(x + 1)$ – the one we started with. We divide this by each of the other denominators in turn to work out what to multiply $A$, $B$ and $C$ by:

For $A$: $(x - 2)^2(x + 1) \div (x - 2)^2 = (x + 1)$, so multiply $A$ by $(x + 1)$
For $B$: $(x - 2)^2(x + 1) \div (x - 2) = (x + 1)(x - 2)$
For $C$: $(x - 2)^2(x + 1) \div (x + 1) = (x - 2)^2$

So we get

$$\frac{x^2 - 9x + 17}{(x-2)^2(x+1)} \equiv \frac{A}{(x-2)^2} + \frac{B}{(x-2)} + \frac{C}{(x+1)} \equiv \frac{A(x+1) + B(x+1)(x-2) + C(x-2)^2}{(x-2)^2(x+1)}$$

so $x^2 - 9x + 17 \equiv A(x + 1) + B(x + 1)(x - 2) + C(x - 2)^2$

Set $x = -1$: $(-1)^2 - 9(-1) + 17 \equiv A(-1 + 1) + B(-1 + 1)(-1 - 2) + C(-1 - 2)^2$
$\Rightarrow 27 = A(0) + B(0) + 9C \Rightarrow C = 3$

Set $x = 2$: $(2)^2 - 9(2) + 17 \equiv A(2 + 1) + B(2 + 1)(2 - 2) + C(2 - 2)^2$
$\Rightarrow 3 = A(3) + B(0) + C(0) \Rightarrow A = 1$

Set $x = 0$: $(0)^2 - 9(0) + 17 \equiv A(0 + 1) + B(0 + 1)(0 - 2) + C(0 - 2)^2$
$\Rightarrow 17 = A(1) + B(-2) + C(4)$

But $A = 1$, $C = 3 \Rightarrow 17 = 1 - 2B + 12 \Rightarrow B = -2$.

*Now learn how to use your knowledge*

# Algebra 2

## Use your knowledge

**Hints**

**1** a) Expand $(1 - 4x)^7$ up to the term in $x^3$.
Hence

b) i) Obtain an approximation to $0.96^7$.

ii) Calculate the percentage error in using this approximation.

c) Show that $(1 + \frac{4}{\sqrt{3}})^7 + (1 - \frac{4}{\sqrt{3}})^7 \approx 226$.

*Don't forget the brackets around $-4x$*
*$(1-4x)^7 = 0.96^7$*

*% error = $\frac{\text{approx-accurate}}{\text{accurate}} \times 100\%$*

*Write down the expansion of $(1+4x)^7$*
*Substitute in for $x$*

**2** Show that the cubic equation $x^3 - x^2 - x - 2 = 0$ has only one real root.

*Factorise it as far as you can*
*How many roots has the quadratic got?*

**3** $f(x) \equiv x^3 + Ax^2 + Bx + 6$ is divisible by $(x + 1)$, and has remainder 60 when divided by $(x - 2)$.

a) Find the values of $A$ and $B$.

b) Factorise $f(x)$ fully.

*Substitute in $x = -1$ and $x = 2$ to get two simultaneous equations*

*You already know one root*

**4** a) Show that $x = -1$ is a solution of the equation $2x^3 - 7x^2 - 5x + 4 = 0$, and find the other two solutions.

b) Hence obtain the solution to the inequality $2x^3 - 7x^2 - 5x + 4 \geq 0$.

*Remember you need a $2x^2$ in the quadratic factor*

*Draw a number line and mark in the roots*
*Check where it is positive*

**5** a) Express $\frac{2x-5}{x-4}$ in the form $A + \frac{B}{x-4}$ where $A$ and $B$ are constants to be determined.

b) Express $\frac{-3x}{2(x-1)(x-4)}$ in the form $\frac{A}{x-1} + \frac{B}{x-4}$, where $A$ and $B$ are constants to be determined.

*Put over a common denominator, and then do partial fractions*
*Write $A = \frac{A}{1}$*

*Make sure you use the common denominator you started with*

Answers on page 86

# Functions

20 minutes

## Test your knowledge

**1** The functions f and g are defined by:

$f(x) = x^2 + 3,\ x \in \mathbb{R}$ and $g(x) = 4x - 5,\ x \in \mathbb{R}$.

a) Sketch the graphs of the two functions $f$ and $g$, showing where they cut the co-ordinate axis.

b) Hence state the range of the functions $f$ and $g$.

c) Explain whether or not (i) $f$ (ii) $g$ have inverse functions, and if so find the inverse.

d) Solve the equation $f(x) = gg(x)$.

**2** The function $f$ is defined by $f(x) = 4x - x^2,\ x \in \mathbb{R}$.

a) Sketch the graph of the function $f$, showing where it cuts the co-ordinate axis.

b) Hence explain why $f(x)$ has no inverse.

The function $h$ is defined by $h(x) = 4x - x^2,\ x \geq c$, is a one–one function.

c) Find the minimum value of c.

d) Find the inverse function, $h^{-1}(x)$.

**3** The graph of $y = f(x)$, as sketched to right, is such that $f(x) = 3$ for $x \leq 1$ and $x \geq 5$.
Sketch the graph of the following showing the new co-ordinates of the points $A$, $B$ and $C$.

a) $y = f(x - 2)$ b) $y = 2f(\frac{1}{2}x)$ c) $y = -f(-x)$.

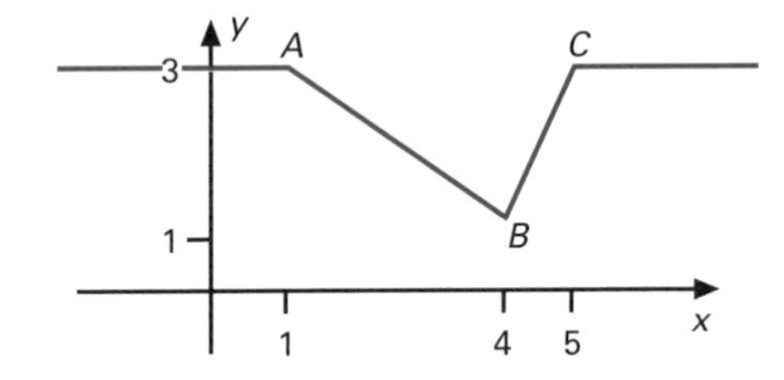


### Answers

1a) $f(x)$: a parabola cutting the $y$-axis at (0, 3)
$g(x)$: a straight line cutting the co-ordinate axis at (0, –5) and (5/4, 0)
b) $f$: $f(x) \geq 3$ or $y \geq 3$ $g$: $g(x) \in \mathbb{R}$ or $y \in \mathbb{R}$
c)i) $f$ is a many–one mapping $\Rightarrow f^{-1}$ is a one–many mapping and is not an inverse function
c)ii) $g$ is a one–one mapping $\Rightarrow g^{-1}$ is also a one–one mapping, and hence a function d) 2, 14
2a) An inverted parabola cutting the co-ordinate axis at (0, 0) and (4, 0).
b) $f(x)$ is a many–one mapping $\Rightarrow f^{-1}$ would be a one–many mapping which is not a function
c) 2 d) $h^{-1}(x) = \sqrt{(4-x)} + 2,\ x \geq 4$
3a) $A(3, 3)$, $B(6, 1)$, $C(7, 3)$ b) $A(2, 6)$, $B(8, 2)$, $C(10, 6)$ c) $A(-1, -3)$, $B(-4, -1)$, $C(-5, -3)$

If you got them all right, skip to page 25

# Functions

**1** A **function** is a rule which converts (i.e. maps) '**input**' values to '**output**' values. A function usually consists of three components: *input*, *rule* and *output*.

**Example 1**

$f(x) = x^2$ is a function.

The input is a value, say $x$. The rule of this function is to **square** the input. Hence the output is $x^2$.

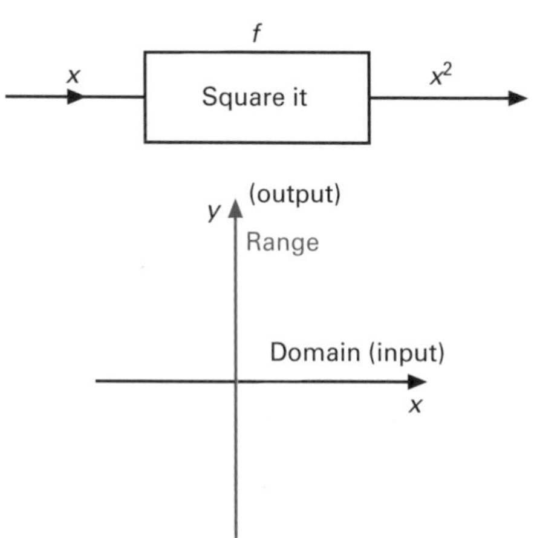


The **domain** is the values that can be put into a function. The **range** is the values that can be output from a function.

When representing a function as a graph, the domain is the $x$-values the function can take and the range are the $y$-values.

For a function to be valid it is allowed to have only **one** output for any given input.

**Example 2**

$f(x) = \pm\sqrt{x}$, is **not** a function because it can have more than one output, e.g. if $x = 9$ (input), then $f(9) = \pm 3$, gives two outputs, which is not possible for a function.

The **domain** of a function is usually given to you in an examination question. For a function, say $f(x) = x^2 + 2$, there are many possible choices of domain, each giving rise to a **range**.

**Example 3**

Give the range of the following functions:

a) $f(x) = x^2 + 2,\ x \in \text{R}$  b) $g(x) = x^2 + 2,\ x \geq 0$  c) $h(x) = x^2 + 2,\ x \leq -1$.

To do this we will sketch the graphs of each of the functions. When sketching the graphs it is important to look at the domain, because this tells us what $x$-values we can draw the graph for. As seen later, $f(x) = x^2 + 2$ is the $y = x^2$ graph moved up two units in the $y$-direction.

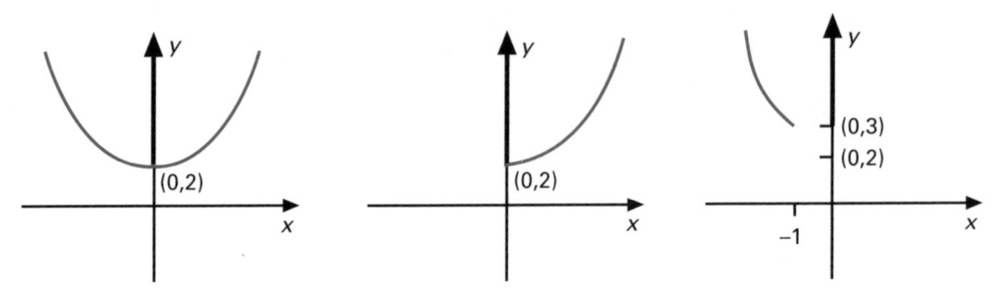


a) Range: $f(x) \geq 2$ or $y \geq 2$ b) Range: $f(x) \geq 2$ or $y \geq 2$ c) Range: $f(x) \geq 3$ or $y \geq 3$

Sometimes it is necessary to restrict the domain so that a function 'makes sense'.

- $f(x) = 1/x$ is **not** defined at $x = 0$, so $x \neq 0$.
- $g(x) = \sqrt{x}$ is **not** possible for negative $x$-values, so $x \geq 0$.

When **two** or more **functions** are **combined** together to form a new function the result is called a **composite function**. With a composite function we appear to work backwards. For $fg(x)$, first '$g$' is applied to $x$, followed by '$f$'.

**Example 4**

For the functions $f(x) = x + 3$, $x \in \mathbb{R}$ and $g(x) = x^2 - 1$, $x \in \mathbb{R}$, find:
a) $fg(x)$ and b) $gf(x)$.

a) Work backwards from $x$: $fg(x)$.
   - Put $g$ into $f$. Every time we see '$x$' in the '$f$' function we replace it by what's in '$g$', i.e. $\mathbf{x^2 - 1}$.
   - This gives $gf(x) = \mathbf{x^2 - 1} + 3 = x^2 + 2$, $x \in \mathbb{R}$.

b) $gf(x) \Rightarrow f$ goes into $g \Rightarrow$ replace $x$ in '$g$' by '$f$', i.e. '$x + 3$'.
   This gives $gf(x) = (\mathbf{x + 3})^2 - 1$, $x \in \mathbb{R}$.

Since $fg(x)$ is a function, it will have a domain. The domain will usually be the domain of the function that $x$ is applied to first, in this case, '$g$'. Since the domain of '$g$' is $x \in \mathbb{R}$, the domain of $fg(x)$ is also $x \in \mathbb{R}$.

Sometimes we must restrict the domain of the first function to make the composite function work.

**Example 5**

For the functions $g(x) = 1/x$, $x \in \mathbb{R}$, $x \neq 0$ and $h(x) = x + 5$, $x \in \mathbb{R}$, $x \neq -5$.
a) Find $gh(x)$, stating its domain, and b) solve the equation $gh(x) = h(x)$.

a) $h$ goes into g $\Rightarrow gh(x) = \dfrac{1}{x+5}$. The domain of $gh(x)$ is $x \in$ R, $x \neq -5$, to prevent the function being undefined at $x = 5$. This means that $x \neq -5$ must be a part of the original function $h(x)$.

b) $\dfrac{1}{x+5} = x+5 \Rightarrow 1 = (x+5)(x+5) \Rightarrow 1 = x^2 + 10x + 25 \Rightarrow x^2 + 10x + 24 = 0$

Factorising gives: $(x + 6)(x + 4) = 0 \Rightarrow x = -6, -4$
You could have used the quadratic equation formula or completed the square to solve the equation in part b).

A **one–one function**, $f(x)$, is a function, that for **every** value of $f(x)$ there exists **only one** corresponding value of $x$.
A **many–one function** $f(x)$, is a function, that for **some** values of $f(x)$ there exists **more than one** corresponding values for $x$.

- $f(x) = x^2$, $x \in$ R, is a many–one function, because when $f(x) = 9$, $x$ could be either 3 or –3.
- $g(x) = 2x + 3$, $x \in$ R is a one–one function because every $y$-value has one corresponding $x$-value.

We can spot whether or not a function is a one–one by drawing its graph. We know that each $y$-value must correspond to only one $x$-value, so if we draw horizontal lines anywhere on the graph, they will only cut the graph once. If any horizontal line cuts the graph more than once then it is a many–one.

$f(x) = x^3$, $x \in$ R

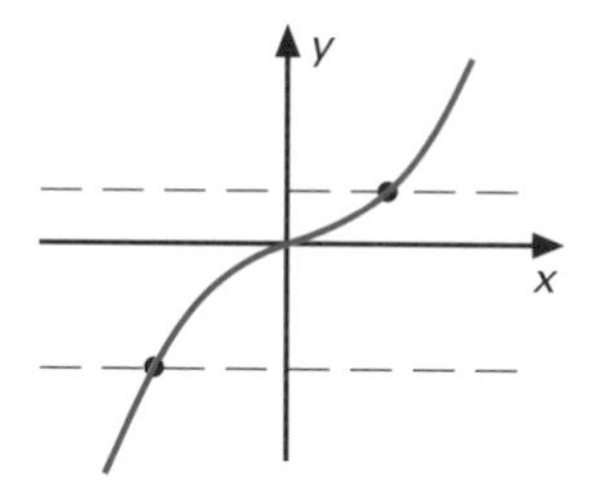


One–one $\Rightarrow f^{-1}(x)$ exists

$f(x) = (x - 1)(x + 2)(x + 1)$, $x \in$ R

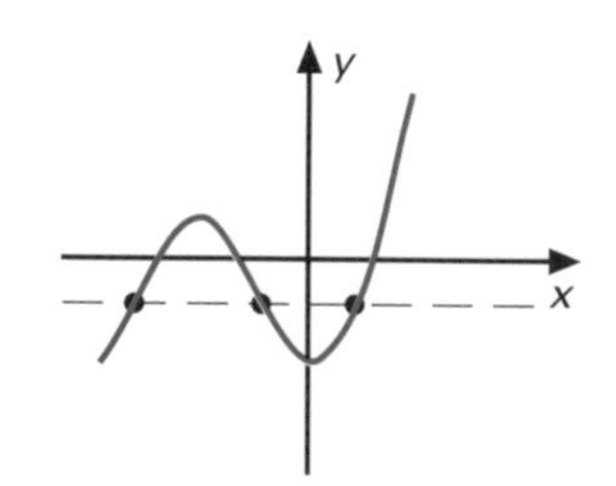


Many–one $\Rightarrow f^{-1}(x)$ does not exist!

An **inverse function** $f^{-1}(x)$ is a function that 'undoes' what $f(x)$ has done. The function $f(x)$ must be a one–one function to have an inverse. The inverse of a many–one function does not exist. This is because the inverse would be a one–many mapping, which has more than one output, and thereby is not a function.

- The domain of a function $f(x)$ becomes the range of the inverse function $f^{-1}(x)$.
- The range of a function $f(x)$ becomes the domain of the inverse function $f^{-1}(x)$.
- The graph of $f^{-1}(x)$ is a reflection of the graph of $f(x)$ in the line $y = x$.

**Example 6**

For the function $f(x) = 7x - 3,\ 1 \leq x \leq 5$

a) State the range.

b) Find the inverse function $f^{-1}(x)$, stating its domain and range.

a) The function is a straight line with positive gradient starting from $x = 1$ and ending at $x = 5$. The smallest value is $f(1) = 7 - 3 = 4$; the largest value is $f(x) = 35 - 3 = 32$.

Hence the range of the function is $4 \leq f(x) \leq 32$.

b) Write down $y = f(x)$: $y = 7x - 3$
- Swap the $x$'s and $y$'s: $x = 7y - 3$
- Rearrange to make $y$ the subject: $x + 3 = 7y \Rightarrow \dfrac{x+3}{7} = y$
- Then write $f^{-1}(x) = y$: $f^{-1}(x) = \dfrac{x+3}{7}$

The domain of $f^{-1}(x)$ is the range of $f(x) \Rightarrow$ domain of $f^{-1}(x)$ is $4 \leq x \leq 32$.
The range of $f^{-1}(x)$ is the domain of $f(x) \Rightarrow$ range of $f^{-1}(x)$ is $1 \leq f^{-1}(x) \leq 5$.

**2** We cannot find the inverse of a many–one function. However, by restricting the domain, we can turn a many–one function into a one–one function, which has an inverse. For quadratic functions, we usually cut the function in half and disregard one half.

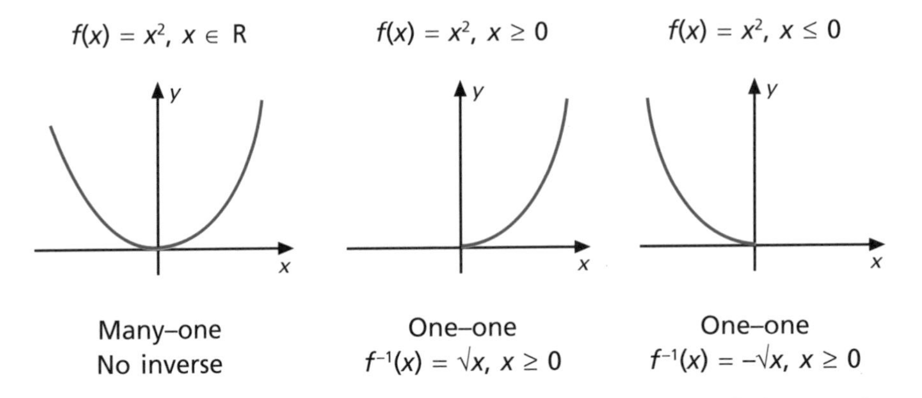


Remember: for a **quadratic** function, the $x$-co-ordinate of the maximum (or minimum) on the curve is found by **averaging** the two roots.

**Example 7**

The function $f$ is defined by $f(x) = x^2 - 10x$, $x \in R$.

a) Sketch the function.

b) Hence explain why $f(x)$ has no inverse.

The function $g$ defined by $g(x) = x^2 - 10x$, $x \geq k$, is a one–one function.

c) Find the minimum value of $k$.

d) Find $g^{-1}(x)$.

**Example 8**

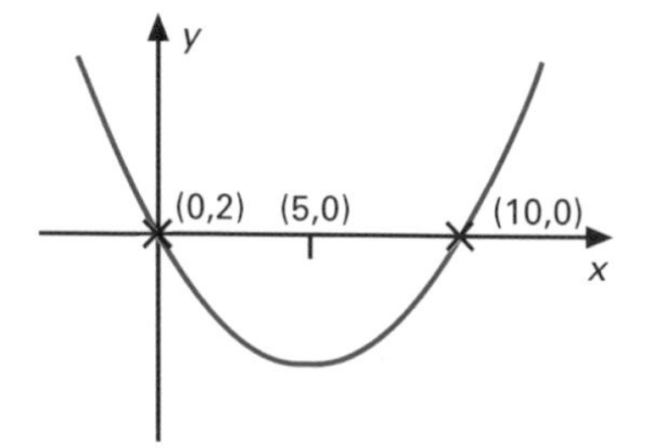


a) $y = x^2 - 10x = x(x - 10) = 0$

$\Rightarrow x = 0$ and $x = 10$ are roots.

This is a happy quadratic, with minimum at $x = \dfrac{0+10}{2} = 5$.

b) The function is a many–one mapping. Hence, the inverse is a one–many mapping, which is not a function.

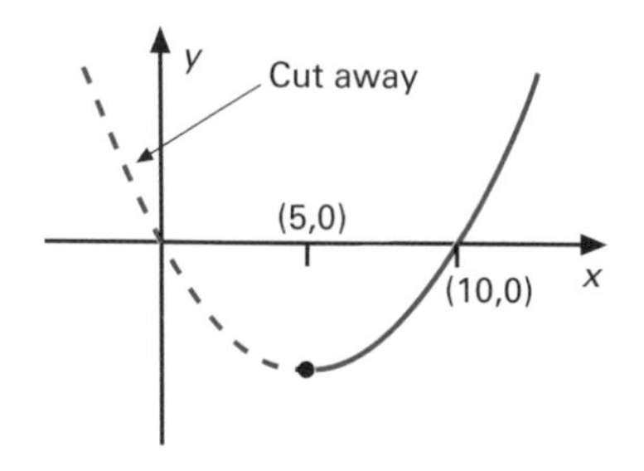


c) $g(x)$ looks like $f(x)$ except the domain of $g(x)$ is different. If we cut out part of the graph of $f(x)$, when $x \leq 5$, then the function will be a one–one function and will have an inverse. Hence the function will be valid for $x \geq 5$. So the minimum value $k = 5$.

d) We need to use the process of completing the square to help us to find the inverse.

- $y = x^2 - 10x$ so $x = y^2 - 10y$
- $x = (y-5)^2 - 25 \Rightarrow x + 25 = (y-5)^2 \Rightarrow \sqrt{(x+25)} = y - 5 \Rightarrow \sqrt{(x+25)} + 5 = y$
- Finally $f^{-1}(x) = \sqrt{(x+25)} + 5$, $x \geq -25$.

**3** You need to learn the basic graphs such as $x^2$, $x^3$, $1/x$, $1/x^2$, $\sin x$, $\cos x$, $\tan x$, $|x|$, $e^x$, $\ln x$, etc. Once you have learnt how to draw these graphs, you can use the method of **transformations of curves** to draw more complicated curves.

If you start with the basic graph of $y = f(x)$, then:

- $y = f(x) + A$: Graph moves up $A$ units
- $y = f(x) - A$: Graph moves down $A$ units
- $y = f(x - A)$: Graph moves to the right $A$ units
- $y = f(x + A)$: Graph moves to the left $A$ units
- $y = f(Ax)$: Stretch scale factor $1/A$ parallel to the $x$-axis
- $y = Af(x)$: Stretch scale factor $A$ parallel to the $y$-axis

- $y = -f(x)$: Reflection of the graph in the $x$-axis
- $y = f(-x)$: Reflection of the graph in the $y$-axis

where $A$ is a positive real number.

**Example 9**

Sketch the graphs of the following:

a) $y = (x + 3)^2$ b) $y = (x - 2)^2 + 3$ c) $y = \dfrac{1}{(x+1)} + 3$.

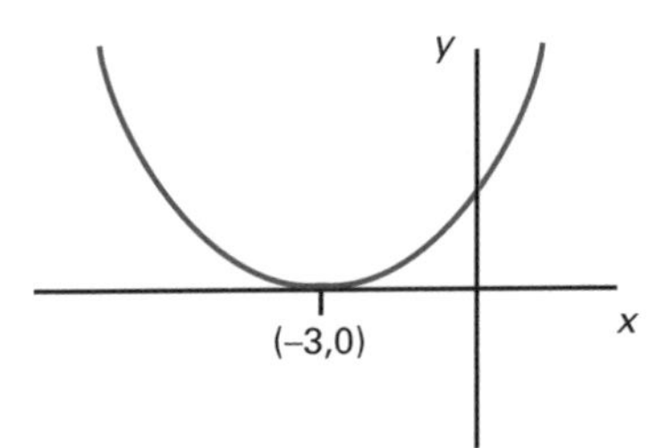


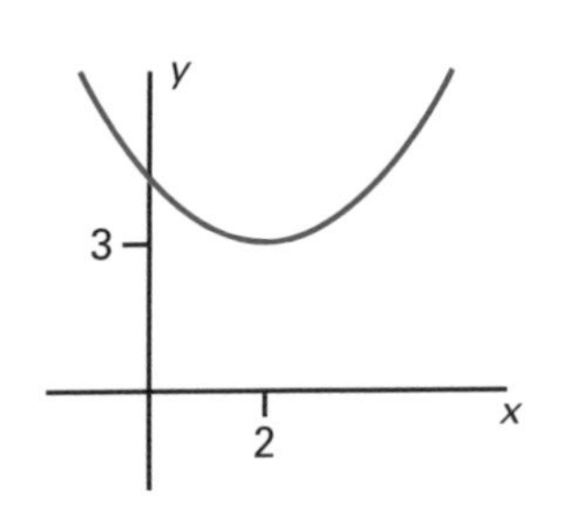


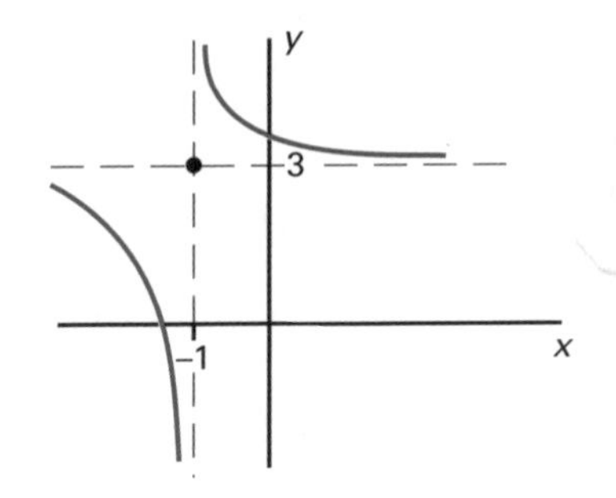


a) This is based on the $y = x^2$ graph. It moves 3 units to the left.

b) The $y = x^2$ graph moves 2 units to the right and 3 units up.

c) The $y = 1/x$ graph moves 1 unit to the left and 3 units up.

**Example 10**

Figure 10 shows the graph of $y = f(x)$.

Sketch on separate axes the graphs with equations:

a) $y = f(2x) - 1$

b) $y = \frac{1}{2}f(x + 3)$.

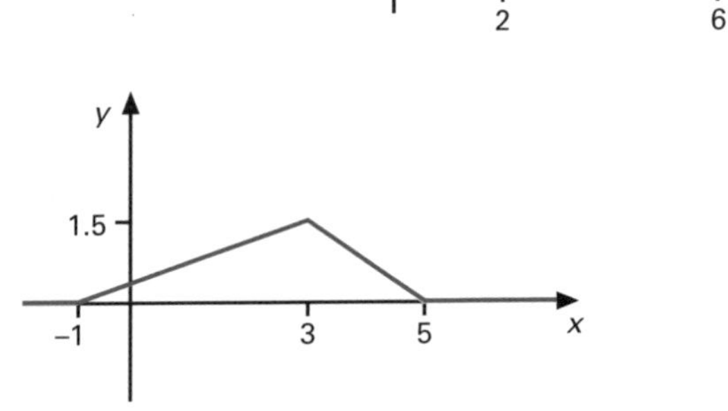


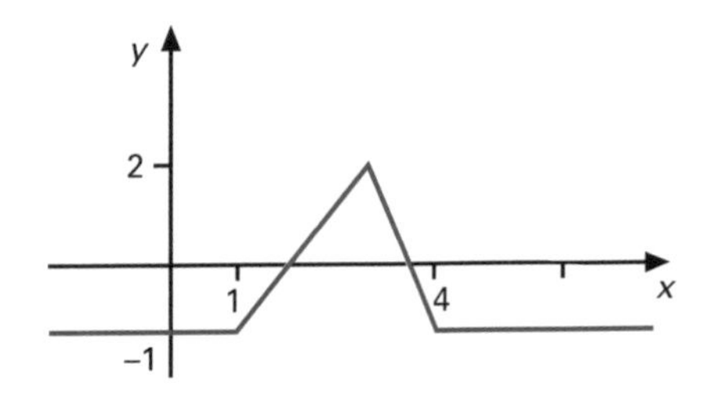


a) This is a stretch SF ½ in the $x$-direction, followed by the graph being moved down 1 unit.

b) This is a stretch SF ½ in the $y$-direction, followed by the graph being moved 3 units to the left.

Now learn how to use your knowledge

# Functions

## Use your knowledge

**Hints**

**1** a) Sketch the graph of $y = a + \dfrac{1}{(x+b)}$ where $a$ and $b$ are positive constants. Write down the equations of the asymptotes.

*The basic curve $y = 1/x$ has been transformed. What does b in brackets tell you?*

*Two asymptotes to find!*

The functions $f$ and $g$ are defined as:

$$f(x) = \frac{1}{x+2} + 1,\ x \geq 0 \quad \text{and} \quad g(x) = \ln x,\ x > 0.$$

b) Sketch the graph of the function $f(x)$, showing where it cuts the co-ordinate axis.

*Look at part (a) for help*
*Remember $f(x)$ is only defined for $x \geq 0$*

c) Hence state the range of $f(x)$.

*Look at your graph in part (b)! What y-values is it defined for?*

The inverse function of $f(x)$ is called $f^{-1}(x)$.

d) State the domain and range of $f^{-1}(x)$.

*Domain $f \equiv$ Range $f^{-1}$*
*Range $f \equiv$ Domain $f^{-1}$*

e) Find the value of $x$, to two decimal places, satisfying the equation $f(x) = g(e^{2x})$.

*ln and e cancel out. You need to solve a quadratic by the formula*

**2** The function g is defined as $g(x) = \dfrac{x-1}{x+3}$, $x \in R$, $x \neq -3$.

a) Express $gg(x)$ in the form $\dfrac{a}{x+b}$, where $a$ and $b$ are constants to be determined.

*Put 'g' into 'g'*
*Replace 'x' in 'g' by $\dfrac{x-1}{x+3}$*

b) Find the inverse function $g^{-1}(x)$.

*Change letters around and make y the subject*

Answers on page 87

# Trigonometry

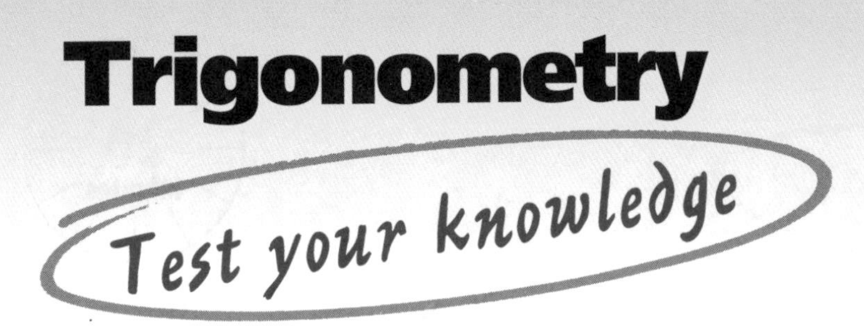

**1** Find the solutions to each of the following equations in the specified ranges:

a) $\tan(x/2) = 0.5;\ -360^0 < x < 360^0$

b) $\sin(x - \pi/9) = -0.5;\ -2\pi < x < 2\pi$

c) $\cos(3x - 90) = -\sqrt{3}/2;\ 0 \le x \le 270°$.

**2** Find the solution to the following equations in the specified ranges:

a) i) $4\sin^3 x - \sin x = 0;\ -\pi \le x \le \pi$ ii) $4\cos^2 x = 3;\ -180° \le x \le 180°$

b) i) $7\sin x + 6\cos^2 x = 8;\ 0 \le x \le 180°$ ii) $\mathrm{cosec}^2 2x = 2\cot 2x;\ -\pi \le x \le \pi$.

**3** Find the solution to the following equations in the specified ranges:

a) $\cos 2x = \cos x;\ -\pi \le x \le \pi$ b) $\tan 2x = 3\tan x;\ 0° \le x \le 360°$

c) $\mathrm{cosec}\, x = 2\sin x;\ -\pi \le x \le \pi$ d) $\sin(x - 30) = 2\cos x\ 0° \le x \le 360°$.

**4** Prove the following identities:

a) $\cos 3x \equiv 4\cos^3 x - 3\cos x$ b) $\sec^2 x\,\mathrm{cosec}^2 x \equiv \sec^2 x + \mathrm{cosec}^2 x$

c) $\sin(x + 30) + \sin(x - 30) \equiv \sqrt{3}\sin x$.

## *Answers*

**1** a) $x = 53.1°, -306.9°$. b) $x = 35\pi/18, -\pi/18, 23\pi/18, -13\pi/18$.
c) $x = 80°, 200°, 100°, 220°$

**2** a) i) $x = 0, \pi, -\pi, \pi/6, 5\pi/6, -\pi/6, -5\pi/6$.
ii) $x = 30°, -30°, 150°, -150°$

b) i) $x = 41.8°, 138.2°, 30°, 150°$ ii) $x = \pi/8, 5\pi/8, -7\pi/8, -3\pi/8$

**3** a) $x = 2\pi/3, -2\pi/3, 0$ b) $x = 0, 180°, 30°, 150°, 210°, 330°, 360°$ c) $x = \pi/4, 3\pi/4, -\pi/4, -3\pi/4$ d) $x = 70.9°, 250.9°$

**4** a) $\mathrm{LHS} = \cos 3x \equiv \cos(2x + x) \equiv \cos 2x\cos x - \sin 2x \sin x \equiv (2\cos^2 x - 1)\cos x - 2\sin x\cos x\sin x$
$\equiv 2\cos^3 x - \cos x - 2\sin^2 x\cos x \equiv 2\cos^3 x - \cos x - 2(1-\cos^2 x)\cos x \equiv 2\cos^3 x - \cos x - 2\cos x + 2\cos^3 x$
$\equiv 4\cos^3 x - 3\cos x = \mathrm{RHS}$

b) $\mathrm{LHS} = \frac{1}{\cos^2 x} \times \frac{1}{\sin^2 x} \equiv \frac{1}{\sin^2 x\cos^2 x}$ $\mathrm{RHS} = \frac{1}{\cos^2 x} + \frac{1}{\sin^2 x} \equiv \frac{\sin^2 x + \cos^2 x}{\cos^2 x\sin^2 x} \equiv -\frac{1}{\cos^2 x\sin^2 x} = \mathrm{LHS}$

c) $\mathrm{LHS} = \sin(x + 30) - \sin(x - 30) \equiv \sin x\cos 30 + \cos x\sin 30 + \sin x\cos 30 - \cos x\sin 30 \equiv 2\sin x\cos 30$
$\equiv 2\sin x\,\sqrt{3}/2 \equiv \sqrt{3}\sin x \equiv \mathrm{RHS}$

*If you got them all right, skip to page 34*

# Trigonometry

**1** If the question is set in radians, you **must** give your answer in **radians**. Questions involving trig functions in calculus or numerical methods are always in radians.

*Remember: convert from degrees to radians by multiplying by $\pi/180$*

If a question requires you to work in radians, you must give your answer in terms of $\pi$ (unless it says something like 'give your answer to two decimal places').

You are often required to solve equations like $\sin(2x - 15^\circ) = \frac{1}{2}$, $-360^\circ < x < 360^\circ$. A **CAST diagram** (see below) is used to do this. This diagram tells you where each of the three trig functions are positive. Example 1 shows the key steps in answering this sort of question.

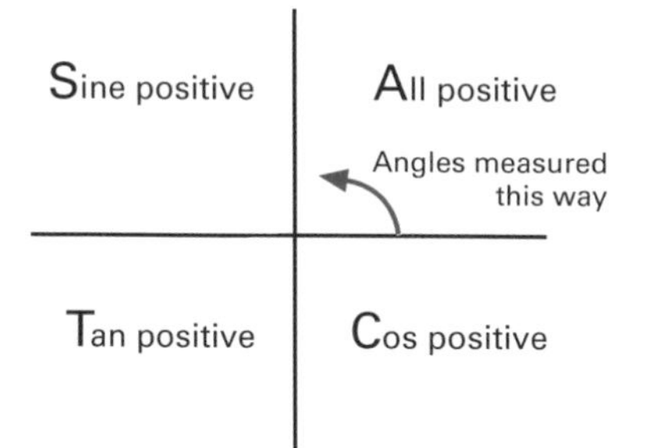


**Example 1**

Find all solutions to the following equations in the specified ranges:

a) $\sin 2x = \frac{1}{2};\ 0^\circ < x^\circ < 360^\circ$

b) $\cos(x - 45) = \frac{1}{\sqrt{2}};\ -270^\circ \leq x^\circ \leq 270^\circ$

c) $\tan(2x - 30) = -1;\ -180^\circ < x^\circ < 180^\circ$.

**Solution**

a) Step 1: Set $2x = u$ $\Rightarrow$ we are solving $\sin u = \frac{1}{2}$

Step 2: Find the values $u$ must be between $\Rightarrow$ since $0^\circ < x^\circ < 360^\circ$, $0^\circ < 2x^\circ < 720^\circ$ (doubling) so $0^\circ < u^\circ < 720^\circ$

Step 3: Decide which quadrants we want $\Rightarrow$ since $\sin u$ is positive, we want 1st and 2nd quadrants

Step 4: Use the calculator to find $\sin^{-1}(\frac{1}{2})$ $\Rightarrow \sin^{-1}(\frac{1}{2}) = 30°$

Step 5: Use the diagram to find 2nd solution $\Rightarrow$ the two solutions are 30° and 150°

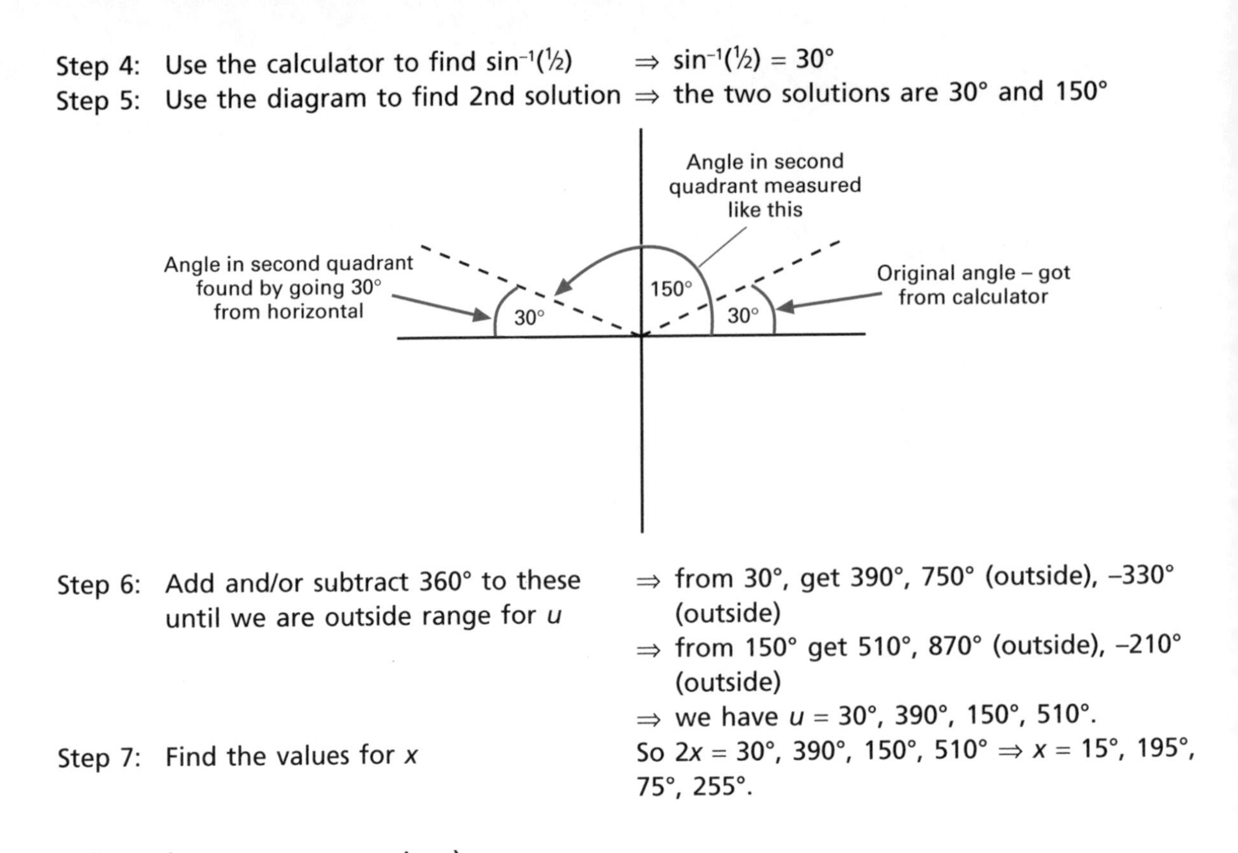


Step 6: Add and/or subtract 360° to these until we are outside range for $u$

$\Rightarrow$ from 30°, get 390°, 750° (outside), –330° (outside)

$\Rightarrow$ from 150° get 510°, 870° (outside), –210° (outside)

$\Rightarrow$ we have $u$ = 30°, 390°, 150°, 510°.

Step 7: Find the values for $x$

So $2x$ = 30°, 390°, 150°, 510° $\Rightarrow x$ = 15°, 195°, 75°, 255°.

b) Follow the same steps as in a):

**1.** $u = x - 45$

**2.** $-270° \le x° \le 270° \Rightarrow -315° \le u° \le 225°$

**3.** cos is positive – so 1st and 4th quadrants

**4.** $\cos^{-1} \frac{1}{\sqrt{2}} = 45°$

**5.** Solutions are 45° and 315°

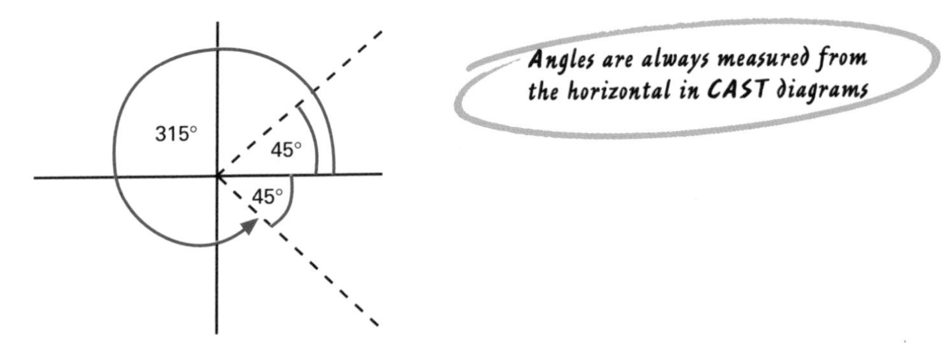


**6.** 45° gives 405° (outside), –315°; 315° gives 675° (outside), –45°

**7.** $x - 45 = u$ = 45°, –315°, 315°, –45° $\Rightarrow x$ = 90°, 270°, 360°, 0°.

c) Same steps again:

**1.** $u = 2x - 30$

**2.** $-180° < x° < 180° \Rightarrow -360° < 2x° < 360° \Rightarrow -390° < 2x - 30 < 330°$

**3.** tan is negative so 2nd and 4th quadrants

**4.** $\tan^{-1}-1 = -45°$

**5.** Solutions are 135° and 315°

**6.** 135° gives 495° (outside), –225°;
315° gives 675° (outside), –45°, –405° (outside)

**7.** $2x - 30 = u = 135°, -225°, 315°, -45° \Rightarrow 2x = 165°, -195°, 345°, -15°$
$\Rightarrow x = 82\frac{1}{2}°, -97\frac{1}{2}°, 172\frac{1}{2}°, -7\frac{1}{2}°$

Sometimes you will need to do this sort of thing in radians. Unless you love radians, you might find it easier to work in degrees and convert at the end (leaving your answer in terms of $\pi$, of course!).

## 2

Sometimes you will meet **polynomial equations** looking like $2\sin^2 x - \sin x - 1 = 0$, i.e. quadratics (or maybe cubics) with a trig function in them instead of just $x$. These are easy: you just factorise (put $y = \sin x$ if it makes it easier for you), get the values for $\sin x$, then get the solutions as above.

**Example 2**

Solve the equation $2\sin^2 x - \sin x - 1 = 0, \quad 0° < x° < 360°$.

**Solution**

We can change this to $2y^2 - y - 1 = 0 \Rightarrow (2y + 1)(y - 1) = 0$
$\Rightarrow y = -\frac{1}{2}$ or $1 \Rightarrow \sin x = -\frac{1}{2}$ or $1$.
$\sin x = -\frac{1}{2} \Rightarrow x = 210°, 330°$; $\sin x = 1 \Rightarrow x = 90°$.

Most equations need you to use some trig definitions. These are the easiest ones:

$\sec x = 1/\cos x$; $\cot x = 1/\tan x$ and $\operatorname{cosec} x = 1/\sin x$

*To remember what sec, cosec and cot are, look at the third letter!*

You also need to know **tanx** = $^{\sin x}/_{\cos x}$ (and so **cotx** = $^{\cos x}/_{\sin x}$).

You also need the **squared ratio identities**. These are probably in your formula book:

A can be any angle: $x$, $4x$, $30°$...

$\sin^2 A + \cos^2 A = 1$; $\tan^2 A + 1 = \sec^2 A$; $1 + \cot^2 A = \text{cosec}^2 A$

Of these, the most important by far is $\sin^2 A + \cos^2 A = 1$; you will find this coming in everywhere!

You use the squared ratio identities if:

i) The equation contains **sin** and **cos**, or **sec** and **tan**, or **cosec** and **cot** (it MUST be of the same angle)

**and** ii) One (or more) of them is **squared**, – e.g. you have sec and $\tan^2$.

Otherwise, you should try putting everything in terms of sin and cos, and attempting to simplify.

**Example 3**

Solve the equations

a) $\sec x = 1 + 2\tan^2 x$; $-180° < x < 180°$  b) $2\cos 2x = \sec 2x$; $0 < x < \pi$

c) $\sin 3x = 4\cos 3x$; $0° < x < 180°$

**Solution**

a) Since this equation contains sec and tan, and one of them is squared, we use $\tan^2 A + 1 = \sec^2 A$:

$\sec x = 1 + 2\tan^2 x = 1 + 2(\sec^2 x - 1) = 1 + 2\sec^2 x - 2 = 2\sec^2 x - 1$

So $\sec x = 2\sec^2 x - 1 \Rightarrow 2\sec^2 x - \sec x - 1 = 0$

$\Rightarrow (2\sec x + 1)(\sec x - 1) = 0 \Rightarrow \sec x = -\frac{1}{2}$ or $1$

To find a solution, we need to change to 'normal' trig functions:

$^1/_{\cos x} = \sec x = -\frac{1}{2}$ or $1 \Rightarrow \cos x = -2$ or $1$

$\cos x = -2$ gives no solutions; $\cos x = 1$ gives $x = 0$

b) Can't use the squared ratio identities ⇒ try putting everything in terms of sin and cos

It doesn't matter that it's $2x$, not $x$ – it would only be a problem if we had, say $\cos x$ and $\sec 2x$

$2\cos 2x = \sec 2x \Rightarrow 2\cos 2x = {}^1/_{\cos 2x} \Rightarrow 2\cos 2x \times \cos 2x = 1$

$\Rightarrow \cos^2 2x = \frac{1}{2} \Rightarrow \cos 2x = \pm\sqrt{\frac{1}{2}}$

So solutions are $2x = 45°, 135°, 225°, 315°$

$\Rightarrow x = 22\frac{1}{2}°, 77\frac{1}{2}°, 112\frac{1}{2}°, 167\frac{1}{2}°$

Converting to radians: $^{\pi}/_8$, $^{3\pi}/_8$, $^{5\pi}/_8$, $^{7\pi}/_8$

Always remember the ± when you take a square root. Missing it out loses lots of marks!

c) Again, we cannot use squared ratio identities – because nothing is squared. Everything is already in terms of sin and cos. In this case, we have to think what else may be useful; tan = $^{\sin}/_{\cos}$ is the only option.

Dividing both sides by cos3$x$: $^{\sin 3x}/_{\cos 3x} = 4 \Rightarrow \tan 3x = 4$
$\Rightarrow 3x = 76°, 256°, 436°$
$\Rightarrow x = 25.3°, 85.3°, 145.3°$.

*Remember this strategy!*

**3** The **addition formulae** are in your formula book:

$\sin(A \pm B) = \sin A\cos B \pm \cos A\sin B$ $\quad \cos(A \pm B) = \cos A\cos B \mp \sin A\sin B$

$$\tan(A \pm B) = \frac{\tan A \pm \tan B}{1 \mp \tan A \tan B}$$

The + and – signs are the **other way up** in some formulae for a **reason**: $\cos(A + B) \equiv \cos A\cos B - \sin A\sin B$

From these, you can derive (by putting B = A) the double-angle identities:

$\sin 2A \equiv 2\sin A\cos A$; $\quad \cos 2A \equiv \cos^2 A - \sin^2 A$; $\quad \tan 2A \equiv \frac{2\tan A}{1 - \tan^2 A}$

$\equiv 2\cos^2 A - 1$

$\equiv 1 - 2\sin^2 A$

*A can be any angle – 30°, x, 3x...*

**If these are not in your formula book, you need to learn them.**

You know to use the addition formulae or the double-angle formulae in an equation when you see something like $\sin(x + 30)$, $\sin 2x$ or $\sin 3x$ (you split this one as $\sin(2x + x)$).

**Example 4**
Find the solution to the following equations in the specified ranges:

a) $5\sin 2x - 4\sin x = 0$; $-180° \leq x \leq 180°$
b) $\sin(x + 60) = 2\cos(x - 45)$; $0 < x < 2\pi$.

**Solution**
a) Since we have a mixture of $x$ and $2x$, use double-angle formulae.

$5\sin 2x - 4\sin x = 0 \Rightarrow 5(2\sin x\cos x) - 4\sin x = 0$
$\Rightarrow 10\sin x\cos x - 4\sin x = 0 \Rightarrow \sin x(10\cos x - 4) = 0$
Hence $\sin x = 0$ or $10\cos x - 4 = 0 \Rightarrow \sin x = 0$ or $\cos x = {}^4/_{10}$
$\sin x = 0 \Rightarrow x = 0°, 180°, -180°$, $\quad \cos x = {}^4/_{10} \Rightarrow x = 66.4°, -66.4°$.

*Never cancel the sinx in a case like this – you lose a solution*

b) Since we do not have the same thing on both sides, we need to expand both:

$\sin(x + 60) \equiv \sin x\cos 60 + \cos x\sin 60 \equiv \frac{1}{2}\sin x + \frac{\sqrt{3}}{2}\cos x$
$\cos(x - 45) \equiv \cos x\cos 45 + \sin x\sin 45 \equiv \frac{1}{\sqrt{2}}\cos x + \frac{1}{\sqrt{2}}\sin x$
So $\frac{1}{2}\sin x + \frac{\sqrt{3}}{2}\cos x \equiv \frac{2}{\sqrt{2}}\cos x + \frac{2}{\sqrt{2}}\sin x$
Collecting terms: $\sin x(\frac{1}{2} - \frac{2}{\sqrt{2}}) = \cos x(\frac{2}{\sqrt{2}} - \frac{\sqrt{3}}{2})$

Dividing: $\tan x = \dfrac{\sin x}{\cos x} = \dfrac{(\frac{2}{\sqrt{2}} - \frac{\sqrt{3}}{2})}{(\frac{1}{2} - \frac{2}{\sqrt{2}})} = -0.5996$

$\tan^{-1}(-0.5996) = -30.9° \Rightarrow x = 149.1°, 329.1°$
$\Rightarrow$ in radians, $x = 2.602, 5.744$.

*You need to know and use:*
$\cos 30 = \sin 60 = \frac{\sqrt{3}}{2}$
(= 0.866...)
$\sin 45 = \cos 45 = \frac{1}{\sqrt{2}}$
(= 0.707...)

**4** **Identities** are things you have to prove; you don't have to find values of $x$. Identities are written with an $\equiv$ sign (not an $=$ sign).

When you are given an identity to prove, there isn't any absolutely fail-safe strategy, but the following is always a good idea:

- Get everything in terms of sin and cos.
- If you have (say) $2x$ on one side and $x$ on the other, use addition/double-angle formulae.
- Put any fractions over a common denominator.
- Try using $\sin^2 + \cos^2 \equiv 1$.
- Use the answer to help you; if it wants just cos, try to find a way to get rid of everything else!

*If you're stuck on an equation, try these tips too!*

Because you are trying to prove something, you shouldn't start out assuming it; it is a good idea to use 'LHS $\equiv$' and 'RHS $\equiv$'. However, you can work on both sides and get them to meet in the middle – just don't work on both at once!

**Example 5**
Prove the following identities:
a) $\tan x + \cot x \equiv 2\operatorname{cosec} 2x$ b) $\sin 3x \equiv 3\sin x - 4\sin^3 x$.

**Solution**
a) Putting everything in terms of sin and cos:

$\text{LHS} \equiv \dfrac{\sin x}{\cos x} + \dfrac{\cos x}{\sin x}$ $\quad \text{RHS} \equiv \dfrac{2}{\sin 2x}$

Using double-angle formulae: $\text{RHS} \equiv \dfrac{2}{2\sin x\cos x} \equiv \dfrac{1}{\sin x\cos x}$

This seems as far as we're likely to get with the RHS, so work on LHS:

$$\text{LHS} \equiv \frac{\sin x}{\cos x} + \frac{\cos x}{\sin x} \equiv \frac{\sin^2 x + \cos^2 x}{\sin x \cos x} \equiv \frac{1}{\sin x \cos x} \equiv \text{RHS}$$

(putting it over a common denominator and using $\sin^2 + \cos^2 \equiv 1$).

b) Everything is already in terms of sin and cos. Since we have $\sin 3x$ on one side and $\sin x$ on the other, we need to use addition formulae:

LHS $\equiv \sin 3x \equiv \sin(2x + x) \equiv \sin 2x \cos x + \cos 2x \sin x$

We still have $2x$ involved, so need to use double-angle formulae. We have to decide which to use for $\cos 2x$; looking at the RHS, we want everything in terms of sin. So use $\cos 2x \equiv 1 - 2\sin^2 x$

LHS $\equiv \sin 2x \cos x + \cos 2x \sin x \equiv (2\sin x \cos x)\cos x + (1 - 2\sin^2 x)\sin x$
$\equiv 2\sin x \cos^2 x + \sin x - 2\sin^3 x$

This looks closer to the answer – everything is in terms of $x$. Comparing what we've got to what we want, we can see that everything except the $\cos^2 x$ looks OK – so we need to get rid of this by using $\sin^2 + \cos^2 = 1$.

So: LHS $\equiv 2\sin x \cos^2 x + \sin x - 2\sin^3 x \equiv 2\sin x(1 - \sin^2 x) + \sin x - 2\sin^3 x$
$\equiv 2\sin x - 2\sin^3 x + \sin x - 2\sin^3 x \equiv 3\sin x - 4\sin^3 x \equiv$ RHS.

*Now learn how to use your knowledge*

20 minutes

*Hints*

**1** a) Prove $\cos 4x \equiv 8\cos^4 x - 8\cos^2 x + 1$.

*$\cos 4x \equiv \cos(2x + 2x)$ Which formula for $\cos 2x$ is the most useful?*

b) Hence solve the equation $\cos 4x = \cos^2 x$; $-360° \leq x \leq 360°$.

*Substitute in for $\cos 4x$ using part a) Let $y = \cos^2 x$. You now have a quadratic*

c) Write down the solutions to the equation $\cos 2x = \cos^2(\tfrac{1}{2}x)$; $-720 \leq x \leq 720°$.

*Let $x = 2A$ Look back at part b)!*

**2** Solve the equation $2\cos x \cos(\tfrac{\pi}{3}) = 2\sin x \sin(\tfrac{\pi}{3}) + 1$; $-\pi \leq x \leq \pi$.

*Take the sinxsin 60 over the other side Try dividing by 2 This looks like a standard formula!*

**3** a) Given that $t = \tan(\tfrac{1}{2}x)$, show that:

i) $\tan x = \dfrac{2t}{1-t^2}$ ii) $\sec x = \dfrac{1+t^2}{1-t^2}$.

*In the formula for $\tan 2A$, put $A = \tfrac{1}{2}x$ Use $\sec^2 x$ and $\tan^2 x$*

b) Hence solve the equation $\sec x = 2\tan x \tan \tfrac{1}{2}x$; $-180 < x < 180$.

*Put everything in terms of t*

**4** Prove the identity $\sec x + \tan x \equiv \dfrac{\cos x}{1-\sin x}$.

*Put LHS in terms of sinx and cosx, and put it over a common denominator Multiply top and bottom by $1 - \sin x$ Use $\sin^2 + \cos^2 = 1$*

Answers on page 87

# Differentiation

## Test your knowledge

**1** Differentiate the following with respect to $x$:

a) $y = x^7$　b) $y = x^2 + 5$　c) $y = 8x^5 - 4x + 5$

d) $y = 3\cos 4x$　e) $y = \ln 5x$　f) $y = -8\tan(x/2)$

g) $y = \dfrac{2x^3 + x^2 - 3}{x^2}$　h) $y = \dfrac{e^{7x} - 3e^{4x} + e^{3x}}{e^{4x}}$　i) $y = \dfrac{4 + 5x^2}{2x}$

j) $y = \dfrac{3x^2 + 5x + 6}{x^{3/2}}$　k) $y = \ln\left(\dfrac{5}{x}\right)$.

**2** Use the chain rule to differentiate the following functions with respect to $x$:

a) $y = e^{x^2 - 5x + 6}$　b) $y = \ln |\sin x|$　c) $y = 3\sin^{-4} x$

d) $y = \dfrac{1}{\cos x}$　e) $y = (6x^2 + 5x + 3)^{11}$　f) $y = \cos^4 x$.

**3** Differentiate the following functions with respect to $x$:

a) $f(x) = x^2 \sin x$　b) $y = \dfrac{e^x}{\cos x}$.　c) $y = x^3(x^2 + 7)^5$

**4** Calculate the gradient of the following functions when $x = 7$:

a) $x = 3t + 1$, $y = e^t$　b) $y = \dfrac{3x - 1}{(x + 3)(x + 1)}$

**5** Find the co-ordinates and determine the nature of the stationary points for the graph of $y = x^3 + 6x^2 - 15x + 12$.

### Answers

**1** a) $7x^6$　b) $2x$　c) $40x^4 - 4$　d) $-12\sin 4x$　e) $1/x$　f) $-4\sec^2(\frac{1}{2}x)$
g) $2 + 6x^{-3}$　h) $3e^{3x} - e^{-x}$　i) $-2x^{-2} + 5/2$
j) $\dfrac{3}{2}x^{-1/2} - \dfrac{5}{2}x^{-3/2} - 9x^{-5/2}$　k) $-\dfrac{1}{x}$

**2** a) $(2x - 5)e^{x^2 - 5x + 6}$　b) $\dfrac{\cos x}{\sin x} = \cot x$　c) $-12\sin^{-5} x \cos x$　d) $\dfrac{\sin x}{\cos^2 x}$　e) $11(12x + 5)(6x^2 + 5x + 3)^{10}$　f) $-4\cos^3 x \sin x$

**3** a) $2x\sin x + x^2\cos x$　b) $\dfrac{e^x(\cos x + \sin x)}{\cos^2 x}$　c) $3x^2(x^2 + 7)^5 + 10x^4(x^2 + 7)^4 = x^2(x^2 + 7)^4(13x^2 + 21)$

**4** a) $e^2/3$　b) $-\dfrac{3}{160}$　**5** Maximum SP at $(-5, 112)$, minimum SP at $(1, 4)$

If you got them all right, skip to page 44

# Differentiation

The notation for the first derivative used by exam boards is $\frac{dy}{dx}$ or $f'(x)$ . The notation for the second derivative is $\frac{d^2y}{dx^2}$ or $f''(x)$. $\frac{d^2y}{dx^2}$ or $f''(x)$ means you differentiate a function f(x) twice.

All exam formulae booklets contain a few standard derivatives, with the exact number depending on the exam board you are dealing with. Make sure you are aware of what is given to you!

**1** Below is a table showing **basic** and **general derivatives** that are obtained using the chain rule. If the chain rule is a nightmare, learn the general derivatives.

| Basic | | General | |
|---|---|---|---|
| $y$ | $\frac{dy}{dx}$ | $y$ | $\frac{dy}{dx}$ |
| $x^n$ | $nx^{n-1}$ | $(ax+b)^n$ | $an(ax+b)^{n-1}$ |
| $e^x$ | $e^x$ | $e^{ax+b}$ | $ae^{ax+b}$ |
| $\ln\lvert x\rvert$ | $\frac{1}{x}$ | $\ln\lvert ax+b\rvert$ | $\frac{a}{ax+b}$ |
| $\sin x$ | $\cos x$ | $\sin(ax+b)$ | $a\cos(ax+b)$ |
| $\cos x$ | $-\sin x$ | $\cos(ax+b)$ | $-a\sin(ax+b)$ |
| $\tan x$ | $\sec^2 x$ | $\tan(ax+b)$ | $a\sec^2(ax+b)$ |

**Example 1** Differentiate the following with respect to $x$:

a) $y = x^5$ b) $y = 7x^4$ c) $y = e^x - 5x + 2$ d) $y = \cos 4x$

e) $y = \tan\left(\frac{3x}{2}\right)$ f) $y = \ln\left(\frac{x}{2}\right)$ g) $y = \frac{x^3 + 3x^4}{x^2}$ h) $y = \frac{2x^7 - x^2}{x^5}$

i) $y = \frac{e^{-3x} + e^{2x}}{e^x}$. j) $y = \frac{2x^2 + 3x - 5}{\sqrt{x}}$ k) $y = \ln\left(\frac{1}{x}\right)$

**Solution** The first six are fairly straightforward:

a) $\frac{dy}{dx} = 5x^4$ b) $\frac{dy}{dx} = 7 \times 4x^3 = 28x^3$ c) $\frac{dy}{dx} = e^x - 5$

d) $\frac{dy}{dx} = -4\sin 4x$ e) $\frac{dy}{dx} = \frac{3}{2}\sec^2\left(\frac{3x}{2}\right)$ f) $\frac{dy}{dx} = \frac{\frac{1}{2}}{\frac{x}{2}} = \frac{1}{x}$.

You must split up the fractions to do the next four :

g) $y = \frac{x^3}{x} + \frac{3x^4}{x^2} = x + 3x^2 \Rightarrow \frac{dy}{dx} = 1 + 6x$

h) $y = \frac{2x^7}{x^5} - \frac{x^2}{x^5} = 2x^2 - x^{-3} \Rightarrow \frac{dy}{dx} = 4x + 3x^{-4} = 4x + \frac{3}{x^4}$

i) $y = \frac{e^{-3x}}{e^x} - \frac{e^{2x}}{e^x} = e^{-4x} - e^x \Rightarrow \frac{dy}{dx} = -4e^{-4x} - e^x$

j) $y = \frac{2x^2}{x^{1/2}} + \frac{3x}{x^{1/2}} - \frac{5}{x^{1/2}} = 2x^{3/2} + 3x^{1/2} - 5x^{-1/2} \Rightarrow \frac{dy}{dx} = 3x^{1/2} + \frac{3}{2}x^{-1/2} + \frac{5}{2}x^{-3/2}$.

We use the subtraction law of logarithms to help us to do the final example:

k) $y = \ln 1 - \ln x = -\ln x \Rightarrow \frac{dy}{dx} = -\frac{1}{x}$.

However there are other differentiation questions where you need to know the rules or special cases.

**2** In your formula book the **chain rule** is stated as follows:

$$\frac{dy}{dx} = \frac{dy}{du} \times \frac{du}{dx} \quad \text{where } y = f(u) \text{ and } u = g(x).$$

This looks scary, but it becomes easier when you look at the next example.

**Example 2** Use the chain rule to find $\frac{dy}{dx}$ for the following:

a) $y = \sqrt{(2 - 3x^2)}$ b) $y = \sin^4 x$.

**Solution**

a) $y = \sqrt{(2 - 3x^2)} = (2 - 3x^2)^{1/2}$

- Let $u$ be what's in the brackets : $u = 2 - 3x^2$
- Substitute $u$ into the original : $y = u^{1/2}$
- Hence $y = u^{1/2}$, where $u = 2 - 3x^2$.

- Differentiate each: $\frac{dy}{du} = \frac{1}{2}u^{-1/2}$ and $\frac{du}{dx} = -6x$
- Use formula $\frac{dy}{dx} = \frac{dy}{du} \times \frac{du}{dx}$: $\frac{dy}{dx} = \left(\frac{1}{2}u^{-1/2}\right) \times (-6x) = -3x\,u^{-1/2}$
- Substituting back: $\frac{dy}{dx} = -3x(2-3x^2)^{-1/2} = \frac{-3x}{(2-3x^2)^{1/2}}$.

b) $y = \sin^4 x = (\sin x)^4$

- $y = u^4$ where $u = \sin x$
- $\frac{dy}{du} = 4u^3$ and $\frac{du}{dx} = \cos x$
- $\frac{dy}{dx} = 4u^3 \cos x = 4\sin^3 x \cos x$.

If you want to save time and effort in an exam, it is good practice to learn the generalised form of the chain rule, as applied in the examples in the table below.

| **Generalised form** | | **Examples** | |
|---|---|---|---|
| $y$ | $\frac{dy}{dx}$ | $y$ | $\frac{dy}{dx}$ |
| $(f(x))^n$ | $n \cdot f'(x) \cdot (f(x))^{n-1}$ | $(4x^2+7)^8$ | $8 \cdot 8x(4x^2+7)^7 = 64x(4x^2+7)^7$ |
| $e^{f(x)}$ | $f'(x)\,e^{f(x)}$ | $e^{-7x^3+5x}$ | $(-21x^2+5)\,e^{-7x^3+5x}$ |
| $\ln f(x)$ | $\frac{f'(x)}{f(x)}$ | $\ln\cos x$ | $\frac{-\sin x}{\cos x} = -\tan x$ |
| $\sin(f(x))$ | $f'(x)\cos(f(x))$ | $\sin(4x^2)$ | $8x\cos(4x^2)$ |
| $\cos(f(x))$ | $-f'(x)\sin(f(x))$ | $\cos(e^{x/2})$ | $-\frac{1}{2}e^{x/2}\sin(e^{x/2})$ |
| $\tan(f(x))$ | $f'(x)\sec^2(f(x))$ | $\tan(1-3x^2)$ | $-6x\sec^2(1-3x^2)$ |
| $\sin^n x$ | $n\sin^{n-1} x\cos x$ | $\sin^5 x$ | $5\sin^4 x\cos x$ |
| $\cos^n x$ | $-n\cos^{n-1} x\sin x$ | $\cos^6 x$ | $-6\cos^5 x\sin x$ |

**3** To use the **product rule** we need to see two separate functions with $x$ in them, which multiply each other. We call one of the products '$u$', the other '$v$'. So $\boldsymbol{y = u.v.}$

The formula for the product rule is found in your formula book:

$$\frac{dy}{dx} = v\frac{du}{dx} + u\frac{dv}{dx}$$

*Find it when you are in an exam!*

There are four ingredients to help us to find $\frac{dy}{dx}$: $u, v, \frac{du}{dx}$ and $\frac{dv}{dx}$.

**Example 3**

Find $\frac{dy}{dx}$ for the function $y = x^2 \cos 3x$.

**Solution** Let $u = x^2$, $v = \cos 3x$

- Differentiate each: $\frac{du}{dx} = 2x \quad \frac{dv}{dx} = -3\sin 3x$
- Use formula: $\frac{dy}{dx} = 2x\cos 3x - 3x^2 \sin 3x$.

To use the **quotient rule** we need one function divided by another. The one at the top is called '$u$', the bottom one '$v$'. So **y = u/v**. It has the formula:

$$\frac{dy}{dx} = \frac{v\frac{du}{dx} - u\frac{dy}{dx}}{v^2}$$

Again, for the quotient rule you need to find the four magic ingredients and insert them into the formula.

**Example 4**

Find the gradient of the curve $y = \frac{e^{2x}}{4x+5}$ when $x = 0$.

**Solution** Functions dividing $\Rightarrow$ use the quotient rule.

- Let $u = e^{2x} \quad v = 4x + 5$
- Differentiate: $\frac{du}{dx} = 2e^{2x} \quad \frac{dv}{dx} = 4$
- Use formula: $\frac{dy}{dx} = \frac{(4x+5)e^{2x} - 4e^{2x}}{(4x+5)^2} = \frac{4xe^{2x} + 5e^{2x} - 4e^{2x}}{(4x+5)^2}$
- Result: $\frac{dy}{dx} = \frac{(4x+1)e^{2x}}{(4x+5)^2}$.

Noting that $\frac{dy}{dx}$ also means the gradient of the curve.

- So when $x = 0$, gradient $= \frac{dy}{dx} = \frac{(4\times 0+1)e^0}{(4\times 0+5)^2} = \frac{1\times 1}{5^2} = \frac{1}{25}$.

**4** In **parametric** equations the variables $x$ and $y$ are expressed in terms of another variable, say '$t$'. The **first derivative** for parametric functions is given by the formula:

$$\frac{dy}{dx} = \frac{dy}{dt} \bigg/ \frac{dx}{dt}$$

**Example 5**

Find the gradient of the curve: $x = 3\sin t,\ y = 2\cos t$, when $t = \pi/4$.

**Solution**

- Differentiate each: $\frac{dx}{dt} = 3\cos t \quad \frac{dy}{dt} = -2\sin t$
- Use formula: $\frac{dy}{dx} = \frac{-2\sin t}{3\cos t} = -\frac{2}{3}\tan t$ because $\frac{\sin t}{\cos t} = \tan t$
- When $t = \pi/4$, $\frac{dy}{dx} = -\frac{2}{3}\tan\left(\frac{\pi}{4}\right) = -\frac{2}{3}$, as required.

Some equations must be expressed as a **partial fraction** before they can be differentiated. Remember to look out for these!

**Example 6**

If $f(x) = \frac{x^2 - 9x + 17}{(x-2)^2(x+1)}$ find $f'(x)$.

**Solution** Express $f(x)$ in partial fractions (see Chapter 2, Algebra 2 for details):

- $f(x) = \frac{1}{(x-2)^2} - \frac{2}{(x-2)} + \frac{3}{(x+1)}$
- Express $f(x)$ in indices form: $f(x) = (x-2)^{-2} - 2(x-2)^{-1} + 3(x+1)^{-1}$
- Differentiate using the chain rule: $f'(x) = -2(x-2)^{-3} + 2(x-2)^{-2} - 3(x+1)^{-2}$.

$f'(x)$ may also be written as: $f'(x) = -\frac{2}{(x-2)^3} + \frac{2}{(x-2)^2} - \frac{3}{(x+1)^2}$.

**5** **Stationary points** (SPs) are the co-ordinates when $\frac{dy}{dx}$ (the gradient of the tangent at that point) is equal to zero. There are three types of stationary point: **maxima**, **minima** and point of **inflection**. We need to find the stationary point(s) and say what type they are.

To begin, we differentiate the function $y = f(x)$, setting the result equal to zero and find the value(s) for $x$ when $\frac{dy}{dx} = 0$. If we are asked for the co-ordinates of the SP, we put the value of $x$ back into the original equation to find the corresponding value of $y$.

To find the type of SP, we can use one of two tests:

Test 1 **Gradient test** Look at the gradient $\frac{dy}{dx}$ either side of the SP, i.e. if $x = 5$ at the SP, then we would work out the value of $\frac{dy}{dx}$ at $x = 4.9$ and $x = 5.1$, say. We find out whether the gradients at these two points are positive or negative, and by looking at the table below, we can ascertain the type of SP.

Test 2 **Second derivative test** Differentiating $\frac{dy}{dx}$ again we get $\frac{d^2y}{dx^2}$. Then we insert the $x$-value of the SP into $\frac{d^2y}{dx^2}$. If the answer is negative, it's a maximum SP; if positive it's a minimum SP. However, if the answer is zero, the SP could still be any of the three SPs, so then we have to perform the gradient test.

| | Maximum | Minimum | Inflection |
|---|---|---|---|
| Shape | | | or |
| Gradient test: value of $\frac{dy}{dx}$ | + 0 − | − 0 + | + 0 + or − 0 − |
| Second derivative test: value of $\frac{d^2y}{dx^2}$ | $\leq 0$ | $\geq 0$ | $= 0$ |

**Example 7**

For the curve $y = \frac{\ln x}{x}$ find a) the exact co-ordinates of the SP. b) the corresponding value of $\frac{d^2y}{dx^2}$. c) State what type of SP it is.

**Solution**

a) Two functions are divided so use the quotient rule.

- Differentiate: $u = \ln x \quad v = x$

$$\frac{du}{dx} = \frac{1}{x} \quad \frac{dv}{dx} = 1 \Rightarrow \frac{dy}{dx} = \frac{1 - \ln x}{x^2}$$

- Set $\frac{dy}{dx} = 0$ to solve for $x$: $\frac{1-\ln x}{x^2} = 0 \quad (\times x^2) \Rightarrow 1 - \ln x = 0$

$$\Rightarrow 1 = \ln x \Rightarrow x = e^1 = e$$

- Find $y$-co-ordinate: $y = \frac{\ln e}{e} = \frac{1}{e} = e^{-1}$ (because $\ln e = 1$.)

Hence the co-ordinates of the SP are $\left(e, \frac{1}{e}\right)$.

b) Quotient rule again.

- Differentiate: $u = 1 - \ln x \qquad v = x^2$

$$\frac{du}{dx} = \frac{-1}{x} \qquad \frac{dv}{dx} = 2x$$

- Find the second derivative: $\frac{d^2y}{dx^2} = \frac{-x - 2x(1-\ln x)}{x^4} = \frac{-3x + 2x\ln x}{x^4} = \frac{-3 + 2\ln x}{x^3}$.

- When $x = e$, $\frac{d^2y}{dx^2} = \frac{-3 + 2\ln e}{e^3} = \frac{-3+2}{e^3} = -\frac{1}{e^3}$.

c) Since $\frac{d^2y}{dx^2} = -\frac{1}{e^3} < 0$ we conclude there is a maximum SP at $\left(e, \frac{1}{e}\right)$.

**Example 8**

Here is a practical problem. A cylindrical open cup is to be made from a thin sheet of precious gold of area $27\pi$ cm$^2$ with no wastage. The volume of the cup is $V$ cm$^3$ and the radius is $r$ cm.

a) Show that $V = \frac{\pi r}{2}(27 - r^2)$.

b) Find the value of $r$ that gives a maximum volume.

c) Prove the volume is a maximum.

d) Work out the maximum volume of the cup.



**Solution** Draw a picture to help. Write down formulae using the information given. We see that the question talks about volumes and surface areas, hence:

For a cylinder: volume $V = \pi r^2 h$ ①

With no lid: surface area = curved bit + base $\Rightarrow 27\pi = 2\pi rh + \pi r^2$ ②

a) Look at the answer: we see that it is in terms of $r$, but the variable $h$ does not appear, implying that it is our mission to eliminate $h$.

- Hence ②/$\pi$ gives $27 = 2rh + r^2 \Rightarrow h = \dfrac{27 - r^2}{2r}$ ③
- Substituting ③ into ①: $V = \pi r^2 \left(\dfrac{27 - r^2}{2r}\right)$
- After cancelling $r$: $V = \dfrac{\pi r}{2}(27 - r^2)$, as required.

b) $V = \dfrac{27\pi r}{2} - \dfrac{\pi r^3}{2}$, from part a) $\Rightarrow \dfrac{dV}{dr} = \dfrac{27\pi}{2} - \dfrac{3\pi r^2}{2} = 0$

- Hence $\dfrac{27\pi}{2} = \dfrac{3\pi r^2}{2} \Rightarrow r^2 = 9 \Rightarrow r = \pm 3$
- Since the radius cannot be negative, we accept $r = 3$ cm.

c) Either **Gradient Test** Look at $\dfrac{dV}{dr}$ when $r = 2.9$ and $r = 3.1$:

- When $r = 2.9$, $\dfrac{dV}{dr} = \dfrac{27\pi}{2} - \dfrac{3\pi(2.9)^2}{2} = 2.780\ldots > 0$
- When $r = 3.1$, $\dfrac{dV}{dr} = \dfrac{27\pi}{2} - \dfrac{3\pi(3.1)^2}{2} = -2.875\ldots < 0$
- Hence

| $r$ | 2.9 | 3 | 3.1 |
|---|---|---|---|
| $\frac{dV}{dr}$ | + / | 0 — | − \ |

$\Rightarrow$ maximum volume when $r = 3$ cm.

or **second derivative test** $\dfrac{d^2V}{dr^2} = -\dfrac{6\pi r}{2} = -3\pi r$

- When $r = 3$, $\dfrac{d^2V}{dr^2} = -9\pi < 0 \Rightarrow$ maximum volume when $r = 3$ cm.

d) Maximum volume occurs when $r = 3$ cm, so substitute $r = 3$ in part (a):

- $V = \dfrac{\pi r}{2}(27 - r^2) = \dfrac{\pi.3}{2}(27 - 9) = 27\pi \text{ cm}^3$.

*Now learn how to use your knowledge*

# Differentiation

## Use your knowledge

Hints

**1** A curve is given by the equation $y = (x + 2)e^{-2x}$.

a) Find $\frac{dy}{dx}$.

*Use the product rule. Factorise $e^{-2x}$ out to make later working easier*

b) Find the co-ordinates of the stationary point $C$, in terms of exponentials.

*Set a) to zero. Find x then y. Remember $e^{-2x} = 0$ has no solutions*

c) Work out $\frac{d^2y}{dx^2}$ at the point C. Hence determine the nature of the stationary point.

*Product rule again! Then substitute x-co-ordinate from b). Apply the test*

**2** If $f(x) = e^{\sin x}$ find:

a) $f'(x)$

*Chain rule*

b) $f''(x)$

*Product rule using a)*

c) the exact values of $\sin x$ for which $f''(x) = 0$.

*Factorise out $e^{-\sin x}$. Use $\cos^2 x = 1 - \sin^2 x$*

**3** A desk tidy with no lid is made from thin card into the form of a cuboid, as shown. The length of the box must be twice the width, where the width is $x$ cm. The volume of the box is 1728 cm³.

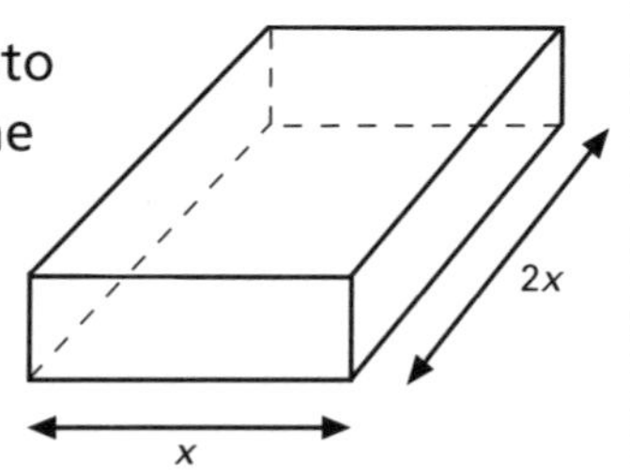


a) Show that the area of card, $A$ cm², used to make the tidy is given by the equation $y = \frac{5184}{x} + 2x^2$.

*Let h = height. Write equations for V and y. Eliminate h*

b) Given that $x$ varies find the value of $x$ to three significant figures, for which $A$ is least and prove that it is a minimum.

*Differentiate y and set to zero. Find x then apply test*

c) Deduce, to three significant figures, the area of the card used to make the base.

*Substitute x value into a)*

Answers on page 88

# Co-ordinate geometry with differentiation

## Test your knowledge

**1** A straight line $l$ passes through $A(2, -2)$, $B(-3, 23)$ and $C(k, 15.5)$ meeting the co-ordinate axis at $D$ and $E$.
a) Calculate the equation of the line $l$ in the from $y = mx + c$.
b) Deduce the value of $k$.
c) Find the length $DE$ to three significant figures.

**2** Show that the triangle $PQR$, where $P(1, 0)$, $Q(3, 4)$ and $R(-3, 2)$, is isosceles.

**3** Calculate the co-ordinates where the line $y = 7x + 28$ intersects the curve $y = 2x^2 + 5x - 12$.

**4** The points $A$, $B$ and $C$ have co-ordinates $(0, 2)$, $(5, -4)$ and $(7, 0)$ respectively. Work out the equation of the straight line $n$ perpendicular to $AB$ and which passes through the midpoint of $BC$. Express your answer in the form $ax + by + c = 0$, where $a$ is a positive integer.

**5** Calculate the equation of a) the tangent and b) the normal to the curve $x = t^3 - 2t$, $y = t^2 + 3$ at the point where $t = 2$. Give your answers to both parts in the form $ax + by + c = 0$, where a is a positive integer.

### Answers

**1a)** $y = -5x + 8$ **b)** $-1.5$ **c)** $8.16$
**2** $PQ = \sqrt{20}$, $QR = \sqrt{40}$, $PR = \sqrt{20} \Rightarrow PQ = PR \neq QR \Rightarrow PQR$ is isosceles.
**3** $(5, 63)$ and $(-4, 0)$
**4** $5x - 6y - 42 = 0$
**5a)** $2x - 5y + 27 = 0$ **b)** $5x + 2y - 34 = 0$

# Co-ordinate geometry with differentiation

## Improve your knowledge

**1** There are three ways you can write an **equation of a line**. They are $\mathbf{y = mx + c}$, $\mathbf{y - y_1 = m(x - x_1)}$, or $\mathbf{ax + by + c = 0}$. The best and easiest formula to use is $y - y_1 = m(x - x_1)$, where $m$ is the gradient of the line, and $(x_1, y_1)$ is a point on the line.

**Think: gradient** and **point** makes **line.**

**Example 1**
Express the equation of the line joining the two points (–6, 4) and (3, 7) in the form $ax + by + c = 0$, where $a$ is a positive integer.

**Solution**: gradient and point $\Rightarrow$ line.

We need the gradient, $m = \dfrac{\text{change in } y}{\text{change in } x} = \dfrac{y_1 - y_2}{x_1 - x_2}$, for points $(x_1, y_1)$ and $(x_2, y_2)$.

So $m = \dfrac{4-7}{-6-3} = \dfrac{-3}{-9} = \dfrac{1}{3}$, and the point = (–6, 4) say.

- Apply the magic formula: $y - 4 = \frac{1}{3}(x - -6)$ $(\times 3)$
- Hence: $3y - 12 = x + 6 \Rightarrow x + 6 - 3y + 12 = 0$
- Gives: $x - 3y + 18 = 0$, as required.

**2** The **distance between two points** is found by:

$\sqrt{(x_1 - x_2)^2 + (y_1 - y_2)^2}$ for points $(x_1, y_1)$ and $(x_2, y_2)$.

**Example 2**
A triangle $ABC$ has co-ordinates $A(4, 5)$, $B(-1, -3)$ and $C(3, -5.5)$.

a) Prove that the angle $A\hat{B}C$ is 90°.

b) Find cos $B\hat{C}A$.

**Solution**

a) Distance AB $= \sqrt{(4--1)^2+(5--3)^2} = \sqrt{25+64} = \sqrt{89}$

Distance BC $= \sqrt{(-1-3)^2+(-3--5.5)^2} = \sqrt{16+6.25} = \sqrt{22.25}$

Distance AC $= \sqrt{(4-3)^2+(5--5.5)^2} = \sqrt{1+110.25} = \sqrt{111.25}$

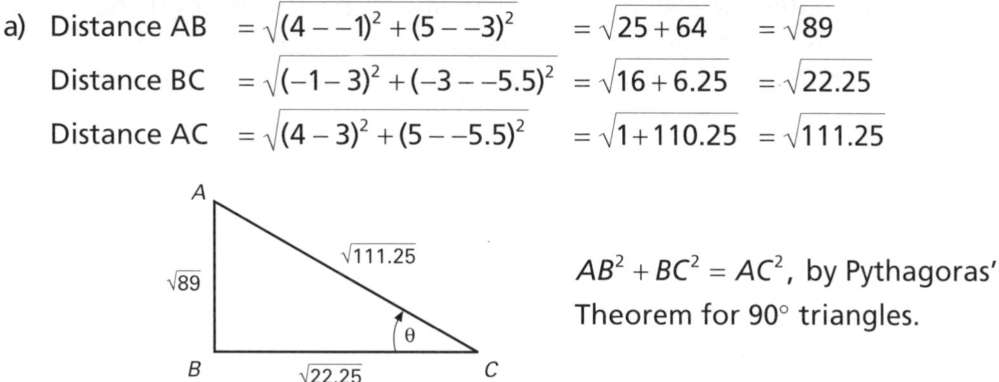


- $\text{LHS} = (\sqrt{89})^2 + (\sqrt{22.25})^2 = 89 + 22.25 = 111.25 = (\sqrt{111.25})^2 = AC^2 = \text{RHS}$
- So: Angle $A\hat{B}C = 90°$, by Pythagoras' Theorem.

b) Angle required is marked as $\theta$

- So $\cos\theta = \frac{\text{Adj}}{\text{Hyp}} = \frac{\sqrt{22.25}}{\sqrt{111.25}} = \sqrt{\frac{22.25}{111.25}} = \sqrt{\frac{1}{5}}$.

The midpoint, M, of the line joining the two points $A(x_1, y_1)$ and $B(x_2, y_2)$ is found by averaging their two co-ordinates. $M = \left(\frac{x_1+x_2}{2}, \frac{y_1+y_2}{2}\right)$

**3** When **lines intersect curves**, we solve the two equations simultaneously.

**Example 3**

The line $y = 2x + 8$ intersects the curve $y = 12x - 2x^2$ at the points $A$ and $B$. Calculate the co-ordinates of the midpoint $M$ of the line $AB$.

**Solution**

Finding the co-ordinates $A$ and $B \Rightarrow$ solve equations simultaneously.

- Hence: $2x + 8 = 12x - 2x^2$, giving $2x^2 - 10x + 8 = 0$
- Factorising: $(2x - 2)(x - 4) = 0 \Rightarrow x = 1$ or $x = 4$
- When $x = 1$, $y = 2(1) + 8 = 10$ when $x = 4$, $y = 2(4) + 8 = 16$
- So $A(1, 10)$ and $B(4, 16)$
- Finally, we find the midpoint: $M = \left(\frac{1+4}{2}, \frac{10+16}{2}\right) \Rightarrow M(2.5, 13)$.

**4** **Parallel lines** have the same gradient i.e. $m_1 = m_2$

Gradients of **perpendicular lines** multiply to make –1 i.e. $m_1 \times m_2 = -1$

**Example 4**
The line $l$ has equation $6x + 4y - 15 = 0$. The line $m$ passes through the point $A(1, 0)$ and is perpendicular to $l$. The line $n$ is parallel to $m$ and passes through the midpoint of $A$ and the point $B(-7, 6)$.

Give the equations of a) the line $m$ and b) the line $n$, in the form $y = mx + c$.

**Solution**
a) We find the gradient of $l$ by writing it in the form of $y = mx + c$.

- Hence: $4y = -6x + 15 \Rightarrow y = \frac{-6}{4}x + \frac{15}{4} = -\frac{3}{2}x + \frac{15}{4}$
- So the gradient of $l$ is $-\frac{3}{2} \Rightarrow$ gradient of $m = \frac{-1}{-3/2} = \frac{2}{3}$ because $l$ is perpendicular to $m$
- Gradient and point $\Rightarrow y - 0 = \frac{2}{3}(x - 1) \Rightarrow y = \frac{2}{3}x - \frac{2}{3}$ which is line $m$.

b) $n$ is parallel to $m$;

- Gradient $n = \frac{2}{3}$ and point $= \left(\frac{1 + -7}{2}, \frac{0 + 6}{2}\right) = (-3, 3)$
- Gradient and point $\Rightarrow y - 3 = \frac{2}{3}(x - -3) \Rightarrow y - 3 = \frac{2}{3}x + 2 \Rightarrow y = \frac{2}{3}x + 5$ which is line $n$.

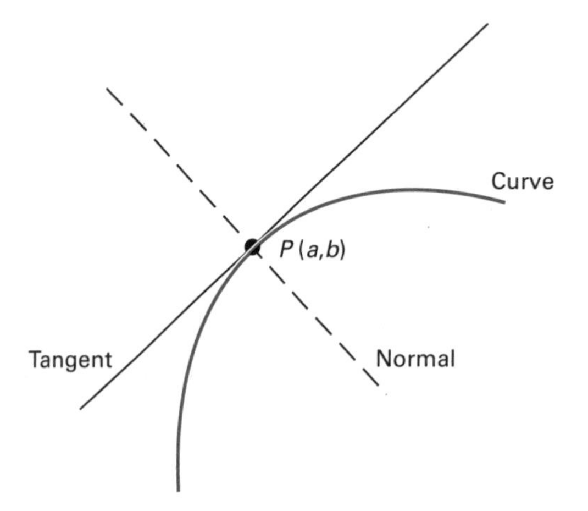


**5** The **tangent T** to the curve is a **line** touching the curve at the point $P$. The **normal N** to the curve at $P$ is a **line perpendicular** to the tangent at $P$.

**Example 5**
Work out (in terms of e) the equations of the tangent and the normal to the curve $y = x^2 e^{2x}$ at the point $Q$, where $x = \frac{1}{2}$.

**Solution**

- Point $Q \Rightarrow y = (\tfrac{1}{2})^2 e^{2.(1/2)} = \tfrac{1}{4}\,e \Rightarrow Q(\tfrac{1}{2}, \tfrac{1}{4}\,e)$
- Differentiate using product rule: $u = x^2$, $v = e^{2x}$

$$\frac{du}{dx} = 2x \qquad \frac{dv}{dx} = 2e^{2x} \qquad \Rightarrow \frac{dy}{dx} = 2xe^{2x} + 2x^2e^{2x}$$

$\frac{dy}{dx}$ is the same as the gradient of the tangent.

- So at $Q$, gradient of $T = \frac{dy}{dx} = 2\left(\frac{1}{2}\right)e^1 + 2\left(\frac{1}{2}\right)^2 e^1 = \frac{3}{2}e$
- Gradient and point $\Rightarrow$ equation of T is $y - \frac{1}{4}e = \frac{3}{2}e\left(x - \frac{1}{2}\right)$.

**Gradient normal** $= \dfrac{\mathbf{-1}}{\textbf{Gradient tangent}}$

- So gradient $N = \dfrac{-1}{\frac{3e}{2}} = \dfrac{-2}{3e}$
- Gradient and point $\Rightarrow$ equation of $N$ is $y - \frac{1}{4}e = \frac{-2}{3e}\left(x - \frac{1}{2}\right)$.

**Example 6**

A curve is defined parametrically by the equations: $x = t^3 - 5t$, $y = t^2$. ①

a) Calculate the equation of i) the tangent $T$ and ii) the normal $N$ to the curve when $t = 2$.

b) Deduce that the tangent intersects the curve when $4t^3 - 7t^2 - 20t + 36 = 0$. ②

c) Calculate the co-ordinates of $Q$, where the tangent cuts the curve again.

**Solution**

a) i) Point: $t = 2 \Rightarrow x = 8 - 10 = -2$ $\quad y = 2^2 = 4$ So $P(-2, 4)$

- Gradient: $\dfrac{dx}{dt} = 3t^2 - 5$ and $\dfrac{dy}{dt} = 2t \Rightarrow \dfrac{dy}{dx} = \dfrac{2t}{3t^2 - 5}$ parametrically
- Gradient at $T$: $\dfrac{dy}{dx} = \dfrac{2(2)}{3(4) - 5} = \dfrac{4}{7}$
- Hence equation of $T$: $y - 4 = \frac{4}{7}(x + 2)$ giving $y = \frac{4}{7}x + \frac{36}{7}$

ii) Gradient $N = \dfrac{-1}{\text{Gradient } T} = \dfrac{-1}{\frac{4}{7}} = -\dfrac{7}{4}$

- Hence equation of $N$: $y - 4 = -\frac{7}{4}(x + 2)$ giving $y = -\frac{7}{4}x + \frac{1}{2}$.

b) This is an intersecting line and curve example:

- Substituting ① into equation for $T$: $t^2 = \frac{4}{7}(t^3 - 5t) + \frac{36}{7}$
- Multiply by 7: $7t^2 = 4t^3 - 20t + 36$
- Hence: $4t^3 - 7t^2 - 20t + 36 = 0$ as required. ②

c) We need to find where the tangent cuts the curve again. To do this we solve equation ②, to find roots (i.e. value for $t$). We know the tangent touches the curve when $t = 2$. This is a solution to ②. Hence $(t - 2)$ is a factor, by the factor theorem.

**Trick** When the tangent **touches** the curve the root is a repeated one. Hence $(t - 2)^2$ is a repeated factor.

- So: $4t^3 - 7t^2 - 20t + 36 = 0 = (t - 2)^2$ (something) $= (t^2 - 4t + 4)$
- So ② $= (t - 2)^2(4t + 9)$. Then $t = 2, 2$ which we already know and $4t + 9 = 0$
- Hence: $t = -\frac{9}{4}$ with $x = \left(\frac{-9}{4}\right)^3 - 5\left(\frac{-9}{4}\right) = -\frac{9}{64}$ and $y = \left(\frac{-9}{4}\right)^2 = \frac{81}{16}$
- Therefore the co-ordinates of intersection with the curve: $Q\left(-\frac{9}{64}, \frac{81}{16}\right)$.

Now learn how to use your knowledge

# Co-ordinate geometry with differentiation

## Use your knowledge

**Hints**

**1** Statto, a couch potato football fan, was asked to design the perfect chair for watching football on his television. Statto got some graph paper and drew a side view of his fantasy chair. He used the scale of one unit in the $x$- and $y$-axis to represent 20 cm.

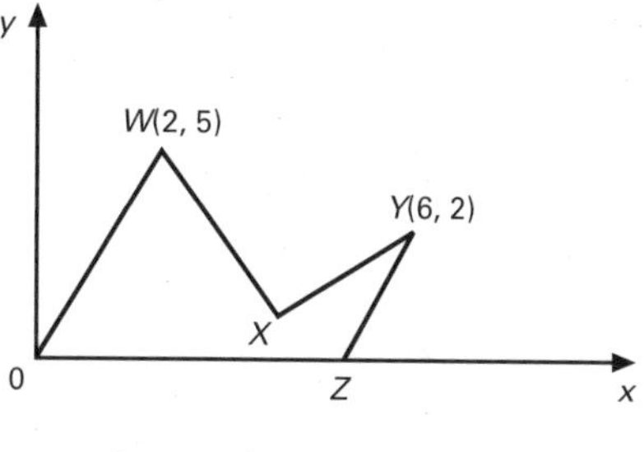


To begin, Statto drew a line from the origin to the point $W$, with co-ordinates (2, 5).

a) Write down the equation of the line $OW$.

*Gradient and point makes line*

Next Statto drew a line $WX$ with gradient –2, as shown in the diagram, which is perpendicular to the line $XY$, with $Y$ having co-ordinates (6, 2).

b) Calculate the equation of the $XY$.

*Work out **WX** first, followed by **WY** using $m_1 m_2 = -1$*

c) Hence calculate the co-ordinates of $X$.

*Equate the lines **WX** and **WY***

Finally the $YZ$ was drawn which is parallel to the line $OW$.

d) Calculate the length the side view occupies on the floor. (Take the $x$-axis as the floor.)

*Work out equation of **YZ***
*Then set $y = 0$ to find where it cuts the x-axis*

**2** A curve is given parametrically by the equations:

$x = 2t,\ y = 6/t$ is defined for $t \in$ R, and $t \neq 0$.

a) Find $dy/dx$ in terms of t.

The normal to the curve is drawn at the point $P$ where $t = 3$.

b) Show that the equation of the normal is $y = 3x - 16$.

The normal meets the curve again at the point $Q$.

c) Calculate the co-ordinates of the point $Q$.

*Use parametric differentiation*
*Find tangent gradient and use $m_1 m_2 = -1$ to find normal*
*Make curve = normal*
*Solve quadratic for t*
*Then find x and y*

Answers on page 89

# Probability and statistics

## Test your knowledge

**1** a) A student shops at Kwikways 30 % of the time, and Safesave the rest of the time. When he goes to each shop, he buys either a can of beans or a pizza. At Kwikways, he buys beans 40 % of the time, but at Safesave he buys beans 65 % of the time.

i) What is the probability that on a randomly selected shopping trip he buys a pizza?

ii) Given that the student bought a pizza, what is the probability he shopped at Kwikways?

b) I throw a fair dice. Event $A$ is getting a prime number. Event $B$ is getting a square number. Event $C$ is getting an even number.

i) Find the probabilities of events $A$, $B$ and $C$.

ii) Show that events $A$ and $B$ are mutually exclusive, and that events $B$ and $C$ are independent.

iii) Are events $A$ and $C$ independent? Justify your answer.

c) Given that $P(A) = 0.6$, $P(B) = 0.3$, and that events $A$ and $B$ are independent, find:

i) $P(A \cap B)$. ii) $P(A|B)$. iii) $P(A \cup B)$.

**2** a) The following table gives some data on the test results (out of 100) obtained by two classes.

Class $A$: mean = 65; standard deviation = 8
Class $B$: mean = 68; standard deviation = 4

i) Which class do you think the person with the highest mark was in? Explain your reasoning.

ii) The teacher marking class $A$'s test had not counted 3 marks on everyone's paper. What are the correct figures for the mean and the standard deviation of the marks in class $A$?

iii) The teacher of class $B$ decided to scale the marks to be out of 50 by dividing all the marks by 2. What will be the new mean and standard deviation be for class $B$?

### Answers

1a) i) 0.425 ii) 0.4235 b) i) $P(A) = \frac{1}{2}$ $P(B) = \frac{1}{3}$ $P(C) = \frac{1}{2}$
ii) $A$ and $B$ cannot happen at once because $A$ only happens if I get 2,3,5, and $B$ only happens if I get 1,4 $P(B \cap C) = P(4) = \frac{1}{6}$ $P(B) \times P(C) = \frac{1}{3} \times \frac{1}{2} = \frac{1}{6}$ so $B$ and $C$ independent iii) $P(A \cap C) = P(2) = \frac{1}{6}$ $P(A) \times P(C) = \frac{1}{2} \times \frac{1}{2} = \frac{1}{4}$ so $A$ and $C$ are not independent c) i) $P(A \cap B) = 0.18$ ii) $P(A|B) = 0.6$ iii) $P(A \cup B) = 0.72$
2a) i) Since the marks in class $A$ are much more variable, they are likely to have the highest mark (and the lowest) ii) Mean = 68; SD = 8 iii) Mean = 34; SD = 2

# Probability and statistics

## Improve your knowledge

20 minutes

**1** You need to be able to recognise and use **probability notation**:

- $P(A \cap B)$ means the probability $A$ and $B$ both happen
- $P(A \cup B)$ means the probability $A$ or $B$ or both happen
- $P(A')$ means the probability $A$ doesn't happen ($= 1 - P(A)$)
- $P(B|A)$ means the probability $B$ happens given $A$ has happened (or the probability of $B$ **conditional** on $A$)

*"A or B" usually means A or B or both in probability.*

From GCSE you should already know how to use tree diagrams. The tree diagrams you will meet at A-level will usually have the second lot of branches depending on the first; see example 1:

*When using a tree diagram, multiply along branches then if you need more than one branch, add them.*

**Example 1**

I have a bag containing 8 red sweets, 3 green sweets and 2 orange sweets. I take 2 sweets out of the bag. What is the probability a) they are the same colour b) they are of different colours?

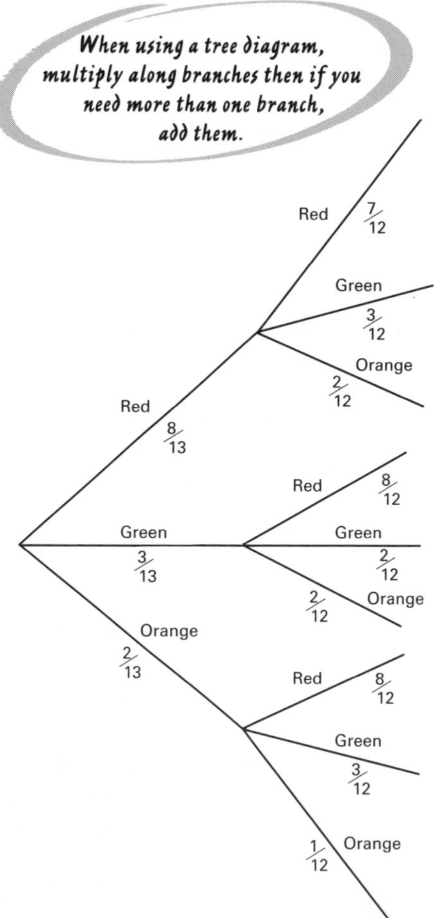


**Solution**

a) To get the same, we need $RR$, $GG$ or $OO$.
$P(RR) = \frac{8}{13} \times \frac{7}{12} = \frac{14}{39}$; $P(GG) = \frac{3}{13} \times \frac{2}{12} = \frac{1}{26}$;
$P(OO) = \frac{2}{13} \times \frac{1}{12} = \frac{1}{78}$
So $P(\text{both the same}) = \frac{14}{39} + \frac{1}{26} + \frac{1}{78} = \frac{16}{39}$

b) Since this is the opposite of both sweets being the same, $P(\text{different}) = 1 - P(\text{same})$
$= 1 - \frac{16}{39} = \frac{23}{39}$.

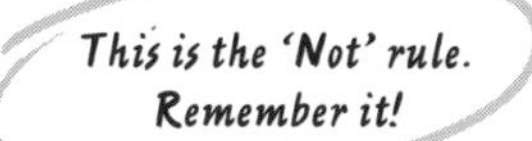

### Formulae

You also need to be able to use the following formulae, which should be in your formula book:

$$P(A \cup B) = P(A) + P(B) - P(A \cap B) \qquad P(B \mid A) = \frac{P(B \cap A)}{P(A)}$$

Example 2 shows you how to use the first formula:

**Example 2**

The probability I have a dessert with my lunch is 0.2. The probability I have a dessert with my dinner is 0.4. On 10 % of days I have desserts with both meals. On what percentage of days do I have:

a) At least one dessert? b) No dessert?

**Solution**

Let dessert with lunch = $L$, dessert with dinner = $D$.

So $P(L) = 0.2$, $P(D) = 0.4$.

We know $P(\text{both}) = 0.1$; 'both' means $L \cap D$, so $P(L \cap D) = 0.1$.

a) We want $P$(at least one dessert) = $P$(lunch or dinner or both) = $P(L \cup D)$.
Using the formula: $P(L \cup D) = P(L) + P(D) - P(L \cap D)$
$= 0.2 + 0.4 - 0.1 = 0.5$, – which is 50% of days.

b) 'No dessert' is the opposite of 'at least one dessert'.
So: $P(\text{no dessert}) = 1 - 0.5 = 0.5$ giving 50 % again.

*You need to know that 'at least one' means 'one or the other or both' and P(at least one) = 1 – P(none)*

### Conditional Probability

Many questions require you to use conditional probability, which is calculated using the second formula above. You know you have to use conditional probability if you see any of the following in a question:

- 'Find the conditional probability that...'
- 'Given that...find the probability that...' or 'Find the probability that...given that...'
- You know something about what's happened already.

When you write $P(B \mid A)$, the thing that you know or that you are given is $A$ and the thing you want to find out is $B$. Example 3 shows how to use it:

**Example 3**

Rena travels to school by bus 40 % of the time, and the rest by bike. When she travels by bus, the probability of her being late is 0.3; when she travels by bike, this probability is 0.2.

a) Find the probability she is late.
b) Given she is late, what is the probability she used the bus?

**Solution**

a) $P(\text{Bus and late}) = 0.4 \times 0.3 = 0.12$
$P(\text{bike and late}) = 0.6 \times 0.2 = 0.12$
$\Rightarrow P(\text{late}) = 0.12 + 0.12 = 0.24.$

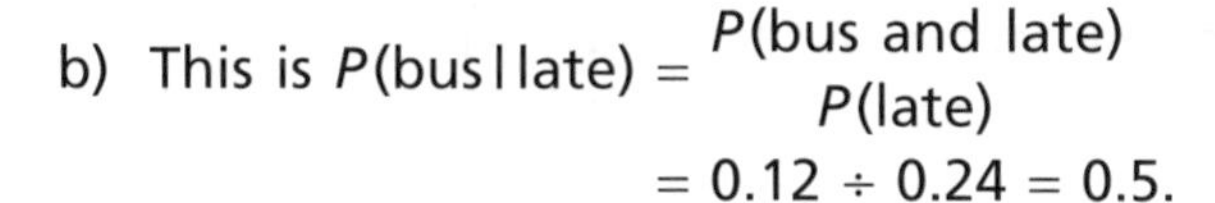

b) This is $P(\text{bus} \mid \text{late}) = \dfrac{P(\text{bus and late})}{P(\text{late})}$
$= 0.12 \div 0.24 = 0.5.$

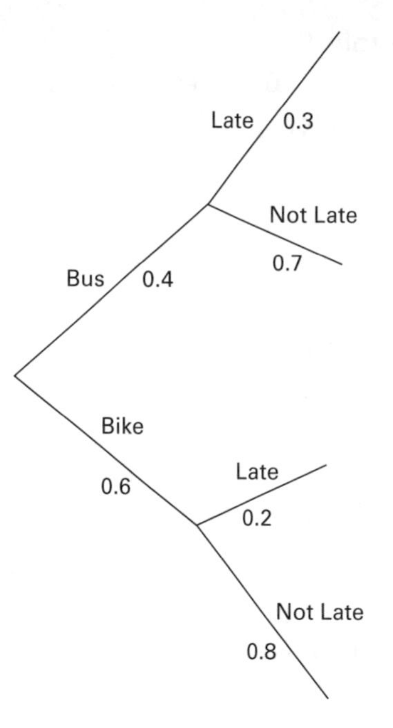


## Independent and Mutually Exclusive

You need to know the meaning of **independent** and **mutually exclusive** events.

- **Independent** events do not affect each other.
**For independent events, $P(A \cap B) = P(A) \times P(B)$.**
- **Mutually exclusive events** can't both happen at once.
**For mutually exclusive events, $P(A \cap B) = 0$.**

Questions can either tell you events are independent/mutually exclusive, when you have to use the formulae above, or they ask you to show events are independent/mutually exclusive, when you have to see if the formulae are true.

**Example 4**

$P(A \cup B) = 0.8$ and $P(A) = 0.5$.
Find $P(B)$ if a) $A$ and $B$ are mutually exclusive, b) $A$ and $B$ are independent.

**Solution**

a) We know $P(A \cap B) = 0$. Since we are given $P(A \cup B)$, it makes sense to use $P(A \cup B) = P(A) + P(B) - P(A \cap B)$.
So: $0.8 = 0.5 + P(B) - 0$ so $P(B) = 0.3$.

b) We know $P(A \cap B) = P(A) \times P(B) = 0.5P(B)$.
Again, use $P(A \cup B) = P(A) + P(B) - P(A \cap B)$.
So: $0.8 = 0.5 + P(B) - 0.5P(B) \Rightarrow 0.3 = P(B) - 0.5P(B) = 0.5P(B)$
$\Rightarrow P(B) = 0.3 \div 0.5 = 0.6.$

**Example 5**

$P(A \cup B) = 0.7$, $P(A) = 0.5$, $P(B) = 0.3$. Investigate whether $A$ and $B$ are independent.

**Solution**

To see whether $A$ and $B$ are independent, we need to check whether $P(A \cap B) = P(A) \times P(B)$ or not. We therefore need to find $P(A \cap B)$. Since we are given $P(A \cup B)$, use $P(A \cup B) = P(A) + P(B) - P(A \cap B)$

So: $0.7 = 0.5 + 0.3 - P(A \cap B) \Rightarrow P(A \cap B) = 0.1$

$P(A) \times P(B) = 0.5 \times 0.3 = 0.15$. $P(A \cap B) = 0.1$

So $P(A) \times P(B) \neq P(A \cap B)$ so $A$ and $B$ are not independent.

## 2

Almost all the **Statistics** on the A-level Pure syllabus you should know from GCSE Higher! In particular, you need to be able to:

- Calculate mode, median, mean, standard deviation, range and interquartile range.
- Draw histograms, scatter diagrams and cumulative frequency diagrams.
- Use scatter diagrams to comment on correlation.
- Use calculated statistics (like mean, median, SD, etc.) to make comments on data.

*Learn how to use the statistical functions on your calculator to find mean and standard deviation*

The mean, median and mode are all 'averages', and the interquartile range and standard deviation are measures of spread – they tell you how spread out the data are.

In addition, you need to know how a change in the data values affects mean, median, SD, etc:

I If **all** the data values go **up** or **down** by a **fixed amount**:

- The **mean, median** and **mode all change** by this amount. So if all the data values were increased by 3, each of these would increase by 3.
- The **standard deviation** and **interquartile range** stay the **same**. These two things do not change because they measure how spread out the data is.

II If **all** the data values are **multiplied** or **divided** by a **fixed** amount:

- Then *everything* is multiplied/divided by that amount. So if all the data values are multiplied by 5, each of the mean, median, mode, standard deviation and interquartile range would be multiplied by 5.

III If only **individual** data values **change**, you have to **recalculate** – but beware:

- A **new** value above the **mean increases** the **mean** (and a value below it decreases the mean).

- A **new** value very **close** to the **mean decreases** the **SD** (and one very far away increases the SD).
- **New** values may **not** affect **median** and **interquartile range** (although they can).

**Example 6**

a) Calculate the mean and standard deviation of 1,2,3,4,5.

b) Hence obtain the mean and standard deviation of:

i) 10,11,12,13,14 ii) 2,4,6,8,10 iii) 32,34,36,38,40

iv) $x + y$, $2x + y$, $3x + y$, $4x + y$, $5x + y$.

**Solution**

a) Mean $= \dfrac{1+2+3+4+5}{5} = 3$ $\quad$ $\text{SD} = \sqrt{\dfrac{1^2+2^2+3^2+4^2+5^2}{5} - 3^2} = \sqrt{2} = 1.414\ldots.$

b) i) We have added 10 to each value, so mean goes up by 10, so is now 13, and SD is unchanged at $\sqrt{2}$.

ii) Have doubled each value, so mean is 6 and SD is $2\sqrt{2}$.

iii) Have doubled each then added 30, so mean is 36 and SD is $2\sqrt{2}$, since not affected by adding 30.

iv) Have multiplied by $x$ and added $y$. So mean is $3x + y$, and SD is $x\sqrt{2}$.

Now learn how to use your knowledge

# Probability and statistics

## Use your knowledge

20 minutes

**Hints**

**1** The data below represents the weights of some students (to the nearest kilogram):

| Weight (kg) | Number of people |
|---|---|
| 40–49 | 2 |
| 50–54 | 10 |
| 55–59 | 30 |
| 60–64 | 40 |
| 65–69 | 15 |
| 70–89 | 3 |

*You need to work out the real lower and upper boundaries*
*Use the class midpoints as your x-values*
*Do you know the real data values?*

a) Find an estimate of the mean and standard deviation of the weights, and explain why it is only an estimate.

b) Draw a histogram to represent this data.

*You need to calculate frequency density = $\frac{\text{Frequency}}{\text{Class width}}$*

c) It was later discovered that the scales used had increased all the students' weights by 10 %. Obtain an estimate for the mean and standard deviation of their true weights.

*Work out what you have to multiply each figure by to get the true weight*
*What happens to mean and SD?*

**2** 0.5 % of the population suffer from a rare disease. There is a test for this disease; it gives a positive result with 99 % of people who have the disease, and with 2 % of people who do not have the disease.

*Draw a tree diagram*
*Don't forget 0.5% is not a probability of 0.5*

a) What is the probability a randomly selected individual will test positive for the disease?

b) Shirley has just tested positive for the disease. What is the probability she actually has it?

*You already know something. What does that tell you about the sort of question it is?*

c) Comment on your answer to part b).

*What proportion of people with positive tests are actually ill?*

**3** a) Explain why $P(A \cap B) + P(A \cap B') = P(A)$.

b) In a certain school, all girls play either hockey or netball or both. Given that $P(\text{netball}|\text{hockey}) = \frac{1}{4}$, $P(\text{netball}) = \frac{11}{20}$ and $P(\text{hockey}) = h$,

i) Explain why $P(N \cap H) + P(N \cap H') + P(N' \cap H) = 1$.

ii) Show $P(\text{just hockey}) = \frac{9}{20}$.

iii) Show that $h = \frac{9}{20} + \frac{1}{4}h$, and hence find the value of $h$.

iv) Find $P(\text{not hockey}|\text{netball})$.

**Hints**

*Think what $A \cap B$ and $A \cap B'$ mean*

*Think what the alternatives are*

*Use the results given in a) with $A$ = netball*

*Use the result given in a) with $A$ = hockey*
*Use the conditional formula to get P(hockey and netball)*

*Use the result in a), with $A$ = netball*
*Work out P(hockey and netball) now you know h*

Answers on page 90

# Integration

**1** Integrate the following with respect to $x$:

a) $\dfrac{1}{(7x-6)^5}$ b) $(4x + 3)^7$ c) $\cos(1 - 3x)$ d) $-\frac{1}{2}\sec^2(x/2)$.

**2** Find:

a) $\displaystyle\int \frac{1}{x^3} + 2\sqrt{x}\,dx$

b) $\displaystyle\int \frac{4}{3x}\,dx$

c) $\displaystyle\int (2x+3)(x-2)\,dx$

d) $\displaystyle\int (e^{-2x}+7)^2\,dx$

e) $\displaystyle\int \frac{2x^2-3x-7}{\sqrt{x}}\,dx$

f) $\displaystyle\int \frac{4x^2+7x-6}{x}\,dx$.

**3** The gradient of a curve is given by the equation $\dfrac{dy}{dx} = \sqrt{x} + \dfrac{1}{x}$, with the curve passing through the point (1, 1). Calculate the equation of the curve.

**4** Find the exact value of:

a) $\displaystyle\int_4^9 (6e^{3x} + 2\sqrt{x})\,dx$

b) $\displaystyle\int_0^{\pi/4} \sin 2x\,dx$.

**5** Find $\displaystyle\int \frac{3x+5}{(x+1)(x+3)}\,dx$.

**6** a) Differentiate $(7 + 2x^2)^{1/2}$ with respect to $x$.

b) Using the result from a), evaluate $\displaystyle\int_1^3 \frac{x}{\sqrt{(7+2x^2)}}\,dx$.

**7** a) Find $\displaystyle\int \cos^2 x\,dx$ b) Evaluate $\displaystyle\int_0^{\pi/3} \cos^2 x\,dx$.

## 8

Find, using the substitutions where given:

a) $\int x(4x^2-5)^7 dx,\ u = 4x^2-5$

b) $\int_0^{\pi/6} \sin^2 x \cdot \cos x dx,\ u = \sin x$

c) $\int \sin^3 x dx$

d) $\int 5x\sqrt{(5x-7)}\, dx.$

## 9

Find:

a) $\int \frac{2x+5}{x^2+5x+8} dx$

b) $\int \frac{x+3}{x^2+6x-11} dx.$

## 10

Find:

a) $\int x \sin x dx$

b) $\int x^2 e^{2x} dx.$

## Answers

**1** a) $-\frac{1}{28}(7x-6)^{-4}+c$ b) $\frac{(4x+3)^8}{32}+c$

c) $-\frac{1}{3}\sin(1-3x)+c$ d) $-\tan\left(\frac{x}{2}\right)+c$

**2** a) $-\frac{1}{2x^2}+\frac{4}{3}x^{3/2}+c$ b) $\frac{4}{3}\ln x+c$

c) $\frac{2x^3}{3}-\frac{x^2}{2}-6x+c$ d) $-\frac{1}{4}e^{-4x}-7e^{-2x}+49x+c$

e) $\frac{4}{5}x^{5/2}-2x^{3/2}+14x^{1/2}+c$ f) $2x^2+7x-6\ln x+c$

**3** $y=\frac{2}{3}x^{3/2}+\ln x+\frac{1}{3}$

**4** a) $2e^{27}-2e^{12}+\frac{76}{3}$ b) $\frac{1}{2}$

**5** $\ln(x+1)+2\ln(x+3)+c=\ln((x+1)(x+3)^2)+c$

**6** a) $\frac{2x}{\sqrt{(7+2x^2)}}$ b) 1

**7** a) $\frac{\sin 2x}{4}+\frac{x}{2}+c$ b) $\frac{\sqrt{3}}{8}+\frac{\pi}{6}$

**8** a) $\frac{(4x^2-5)^8}{64}+c$ b) $\frac{1}{24}$

c) $\frac{1}{3}\cos^3 x-\cos x+c$ d) $\frac{2}{25}(5x-7)^{5/2}+\frac{14}{3}(5x-7)^{3/2}+c$

**9** a) $\ln(x^2+5x+8)+c$ b) $\frac{1}{2}\ln(x^2+6x-11)+c$

**10** a) $-x\cos x+\sin x+c$ b) $\frac{1}{2}x^2e^{2x}-\frac{1}{2}xe^{2x}+\frac{1}{4}e^{2x}+c$

*If you got them all right, skip to page 72*

# Integration

## Improve your knowledge

**Integration** is the reverse process of differentiation. Exam formulae booklets give you some standard integrals and rules, but it is your job to learn the rest.

**1** A lot of valuable time can be saved when we have to integrate basic functions which are expressed in a more general **linear** form. For example, instead of integrating $e^x$ or sin x, we have to integrate $e^{2x+5}$ or sin (a$x$ + b). We call these **standard integrals** and these must be learnt.

| General case | | An example | |
|---|---|---|---|
| $f(x)$ | $\int f(x)\,dx$ | $f(x)$ | $\int f(x)\,dx$ |
| $(ax+b)^n$ | $(ax+b)^{n+1}/[a(n+1)]$ | $(\mathbf{7x-3})^{11}$ | $(1/84)(7x-3)^{12}+c$ |
| $e^{ax+b}$ | $(1/a)e^{ax+b}$ | $e^{\mathbf{1-2x}}$ | $(-1/2)e^{1-2x}+c$ |
| $1/(ax+b)$ | $(1/a)\ln(ax+b)$ | $1/(\mathbf{3-4x})$ | $(-1/4)\ln(3-4x)+c$ |
| $\sin(ax+b)$ | $-(1/a)\cos(ax+b)$ | $\sin \mathbf{2x}$ | $-(1/2)\cos 2x+c$ |
| $\cos(ax+b)$ | $(1/a)\sin(ax+b)$ | $\cos(\mathbf{1-5x})$ | $(-1/5)\sin(1-5x)+c$ |
| $\sec^2(ax+b)$ | $(1/a)\tan(ax+b)$ | $\sec^2(\mathbf{-4x+3})$ | $(-1/4)\tan(-4x+3)+c$ |

It is important to note that the standard integrals above only work if the term **in bold** is **linear**. This means that the highest power of $x$ is 1, so terms like $2x-3$, $1-5x$ or $7x$ are allowed but not $2x^2+3$ nor $x^3-7x-2$.

**2** Sometimes basic algebraic manipulation is needed before we can integrate:

a) $\int 2/(7x)\,dx$
$= (2/7)\int 1/x\,dx$
$= (2/7)\ln x + c$
$\Rightarrow \int a/x\,dx = a\ln x + c$

b) $\int (2x^2+3)/x\,dx$
$= \int (2x^2/x) + (3/x)\,dx$
$= \int 2x + (3/x)\,dx$
$= x^2 + 3\ln x + c$

*Watch out for ln integrals*

c) $\int (2x-1)(4x+7)\,dx$
$= \int 8x^2 + 10x - 7\,dx$
$= (8/3)x^3 + 5x^2 - 7x + c$

d) $\int (e^{2x}-3)^2\,dx$
$= \int (e^{2x}-3)(e^{2x}-3)\,dx$
$= \int e^{4x} - 6e^{2x} + 9\,dx$
$= (1/4)e^{4x} - 3e^{2x} + 9x + c.$

*Can you easily multiply out?*

Do not get worried about the $\int$ and the d$x$; these just tell you to **integrate**! For indefinite integration (that's integration without limits) it is important that you include a constant + **c** at the end of your answer.

**3** In some questions you will need to find the value of $c$, the constant of integration.

**Example 1**
The gradient of the curve is given by the equation d$y$/d$x$ = $3x^2 + 4x + 7$, and the curve passes through the point (1, 12). Calculate the equation of the curve.

**Solution**
We are given the equation for the gradient, d$y$/d$x$; **we want** the curve '$y$ = .....'. To go from d$y$/d$x$ back to '$y$ = .....', we integrate:

- $y = \int 3x^2 + 4x + 7 \, dx = x^3 + 2x^2 + 7x + c$ and we know when $x = 1$, $y = 12$
- So: $12 = 1^3 + 2(1)^2 + 7(1) + c \Rightarrow 12 = 10 + c \Rightarrow c = 2$
- Hence the equation of the curve is: $y = x^3 + 2x^2 + 7x + 2$.

**4** **Definite integration** involves limits and is usually associated with finding the areas between curves.

**Example 2**
Evaluate the following integral in terms of e: $\int_1^4 (4e^{2x} + 3x^{1/2})\, dx$.

**Solution** Integration as usual, do not use + $c$

- Put integrand in square brackets with limits: $\int_1^4 (4e^{2x} + 3x^{1/2})dx = [2e^{2x} + 2x^{3/2}]_1^4$
- Put in the limits and work out 'Top expression – bottom expression': $= (2e^{2(4)} + 2(4)^{3/2}) - (2e^{2(1)} + 2(1)^{3/2})$
- Evaluate: $= 2e^8 - 2e^2 + 14$.

**5** Some exam questions will ask you to integrate expressions which need to be expressed in **partial fractions** before they can even be integrated.

**Example 3**
Show that $\displaystyle\int_{2.5}^{5} \frac{7x^2 - 14x + 6}{(2x-1)(x-1)^2} dx = \ln 24 - \frac{5}{12}$.

**Solution** Looking at Algebra 2 Section 3 we can express the integral as a partial fraction.

- Hence: $\dfrac{7x^2 - 14x + 6}{(2x-1)(x-1)^2} = \dfrac{A}{2x-1} + \dfrac{B}{x-1} + \dfrac{C}{(x-1)^2}$
- This gives: $7x^2 - 14x + 6 = A(x-1)^2 + B(2x-1)(x-1) + C(2x-1)$
- Let $x = 1$: $7 - 14 + 6 = 0 + 0 + C(1) \Rightarrow C = -1$
- Let $x = 0.5$: $7/4 - 7 + 6 = \frac{1}{4}\text{A} \Rightarrow \frac{3}{4} = \frac{1}{4}\text{A} \Rightarrow \text{A} = 3$
- Try $x = 0$: $0 - 0 + 6 = A(1)^2 + B(-1)(-1) + C(-1) \Rightarrow 6 = A + B - C$
- Substituting for $A$ and $C$: $6 = 3 + B + 1 \Rightarrow B = 2$
- So: $\displaystyle\int_{2.5}^{5} \frac{7x^2 - 14x + 6}{(2x-1)(x-1)^2}\,\mathrm{d}x = \int_{2.5}^{5} \left(\frac{3}{2x-1} + \frac{2}{x-1} - \frac{1}{(x-1)^2}\right)\mathrm{d}x.$

Let's try to integrate. The first two terms are ln integrals because they are of the form of (number)/($ax + b$), and the final term can be expressed in the form $(ax + b)^n$.

- So: $\displaystyle\int_{2.5}^{5} \left(\frac{3}{2x-1} + \frac{2}{x-1} - (x-1)^{-2}\right)\mathrm{d}x$

$$= \left[\frac{3}{2}\ln|2x-1| + 2\ln|x-1| - \frac{(x-1)^{-1}}{(-1)}\right]_{2.5}^{5}$$

$$= \left[\frac{3}{2}\ln|2x-1| + 2\ln|x-1| + \frac{1}{x-1}\right]_{2.5}^{5}$$

$$= \left(\frac{3}{2}\ln 9 + 2\ln 4 + 1/4\right) - \left(\frac{3}{2}\ln 4 + 2\ln\frac{3}{2} + \frac{1}{(3/2)}\right)$$

$$= \left(\ln 9^{3/2} + \ln 4^2 + 1/4\right) - \left(\ln 4^{3/2} + \ln(3/2)^2 + 2/3\right)$$

$$= \left(\ln 27 + \ln 16 + 1/4\right) - \left(\ln 8 + \ln(9/4) + 2/3\right)$$

$$= \left(\ln 27 + \ln 16 + 1/4\right) - \left(\ln 8 + \ln 9 - \ln 4 + 2/3\right)$$

$$= \ln 27 + \ln 16 - \ln 8 - \ln 9 + \ln 4 + \tfrac{1}{4} - 2/3$$

$$= \ln\left(\frac{27 \times 16 \times 4}{8 \times 9}\right) - \frac{5}{12} = \ln\left(\frac{1728}{72}\right) - \frac{5}{12}$$

$$= \ln 24 - \frac{5}{12}, \text{ as required.}$$

To do definite integration with partial fractions, you can see that it is important to be familiar with the rules of logarithms.

**6** **Integration is the reverse of differentiation**. Some questions may give you something to differentiate in the first part and then ask you to integrate an expression which looks similar to the answer of the first part.

**Example 4**

a) Differentiate $(9 + 2x^3)^{1/2}$ with respect to $x$.

b) Using the result from a) evaluate $\displaystyle\int_0^2 \frac{x^2}{\sqrt{(9+2x^3)}}\,dx$.

**Solution**

a) We use the quick version of the chain rule to differentiate:

- $\dfrac{dy}{dx} = \tfrac{1}{2}(6x^2)(9+2x^3)^{-1/2} = 3x^2(9+2x^3)^{-1/2} = \dfrac{3x^2}{\sqrt{(9+2x^3)}}$.

b) Surprise! Surprise! The answer to part a) looks like the integral in part b), except for the number 3.

- $\displaystyle\int_0^2 \frac{x^2}{\sqrt{(9+2x^3)}}\,dx = \frac{1}{3}\int_0^2 \frac{3x^2}{\sqrt{(9+2x^3)}}\,dx$

  The ⅓ is there to make the integral multiply out to what's asked in b).

- Using the fact that integration is the reverse of differentiation; we get

  $\displaystyle\frac{1}{3}\int_0^2 \frac{3x^2}{\sqrt{9+2x^3}}\,dx = \frac{1}{3}\left[(9+2x^3)^{1/2}\right]_0^2 = \frac{1}{3}\left((9+16)^{1/2} - 9^{1/2}\right) = \frac{1}{3}(5-3) = \frac{2}{3}$.

**7** When doing integration it is important that you are aware of these **three trigonometric identities** from Chapter 5:

$\sin^2 x + \cos^2 x = 1$ $\quad$ $\cos 2x = 2\cos^2 x - 1$ $\quad$ $\cos 2x = 1 - 2\sin^2 x$

**Example 5** (a classic integral) Find $\int \sin^2 x \, dx$.

**Solution**

This seems easy, but it is not! The **only way** you can integrate $\sin^2 x$ (or even $\cos^2 x$) is by using the double-angle formula for cos. There's no other way, so **learn** it!

There are three versions of the double-angle formula for cos. Since we are integrating $\sin^2 x$, we choose the version only containing sine squareds.

- Choose the appropriate identity: $\cos 2x = 1 - 2\sin^2 x$
- Rearrange to make $\sin^2 x$ the subject: $\sin^2 x = \frac{1}{2}(1 - \cos 2x)$
- Substituting for $\sin^2 x$: $\int \frac{1}{2}(1 - \cos 2x)\,dx$
- Then integrate: $= \frac{1}{2}(x - \frac{1}{2}\sin 2x) + c$
- Simplifying: $= \frac{1}{2}x - \frac{1}{4}\sin 2x + c$.

**8** **Integration by substitution** is a way in which a difficult integral can be broken down into an easier form which can easily be integrated using the standard methods. The question is usually posed in one variable, say $x$, and by making a substitution, we completely change the integral to be expressed into a different variable, say $u$. You are usually given the substitutions in examination questions.

**Example 6**

Find $\int x^2(2x^3 + 7)^5\,dx$ using the substitution $u = 2x^3 + 7$. ①

**Solution**

- Differentiate the substitution: $\frac{du}{dx} = 6x^2$
- Rearrange the differential: $\frac{du}{6} = x^2 dx$ ②

We hate and want to get rid of the $x$'s but we now love the $u$'s. Using equations ① and ② we can eliminate the $x$'s.

- Make the substitutions ① and ②: $\int x^2(2x^3+7)^5\,dx = \int u^5.\frac{du}{6}$

  Now look: no $x$'s

- Integrate with respect to $u$: $= \frac{1}{6}\left(\frac{u^6}{6}\right) + c$
- Substitute for $u$: $= \frac{1}{36}(2x^3+7)^7 + c$.

For definite integrals, we need to change the limits when we use integration with substitution. This is seen in the next example.

**Example 7**

Evaluate $\int_{\pi/6}^{\pi/2} \sin^3 x \cos x\,dx$, using the substitution $u = \sin x$.

**Solution** Start as before:

- Differentiate substitution: $\frac{du}{dx} = \cos x$

- Rearrange the differential: $du = \cos x\, dx$
- Change limits from $x$ to u using the substitution formula: When $x = \pi/2$, $u = \sin(\pi/2) = 1$; When $x = \pi/6$, $u = \sin(\pi/6) = ½$
- Make the substitutions remembering to change limits: $= \int_{0.5}^{1} u^3 du$
- Integrate with respect to $u$: $= ¼ \left[u^4\right]_{0.5}^{1}$
- Evaluate: $= ¼ (1 - 1/16) = 15/64$.

When applying definite integrals for trigonometric functions, the limits must always be in radians.

**Example 8** (Another classic integral) Find $\int \cos^3 x\, dx$.

**Solution**

The **only way** you can integrate $\cos^3$ (or even $\sin^3$) is by using the trigonometric identity $\cos^2 x + \sin^2 x = 1$, followed by integration by substitution.

- Split up integral: $\int \cos^3 x\, dx = \int \cos^2 x \cos x\, dx$
- Use the identity $\cos^2 x = 1 - \sin^2 x$: $= \int (1 - \sin^2 x) \cos x\, dx$

At this stage we need to make a substitution. Let $u = \sin x$, because sine is raised to the highest power and sine when differentiated goes to cos, which is also a part of the integral.

- Hence: $u = \sin x \Rightarrow \dfrac{du}{dx} = \cos x \Rightarrow du = \cos x\, dx$
- Substituting the above into the integral: $\int (1 - u^2)\, du = u - (1/3)u^3 + c$
- Substituting back for $u$: $\int \cos^3 x\, dx = \sin x - (1/3) \sin^3 x + c$.

**Example 9**

Find $\int 2x \sqrt{(2x - 5}\ dx$ using the substitutions $u = 2x - 5$.

**Solution**

To begin, $\dfrac{du}{dx} = 2 \Rightarrow 2dx = du$

- Looking at the integral we know: a) $2\, dx = du$ and b) $\sqrt{(2x - 5)} = \sqrt{u} = u^{1/2}$, but we still have $x$ to change into terms of $u$.

  We use $u = 2x - 5$, which becomes $x = ½ (u + 5)$ after rearrangement.

- Now we can make an effective substitution:

$$\int 2x\sqrt{(2x-5)}\,dx = \int \tfrac{1}{2}(u+5)(u)^{1/2}du = \tfrac{1}{2}\int u^{1/2}(u+5)\,du$$
$$= \tfrac{1}{2}\int u^{3/2}+5u^{1/2}du = \tfrac{1}{2}\left\{(2/5)u^{5/2}+(10/3)u^{3/2}\right\}+c = (1/5)u^{5/2}+(5/3)u^{3/2}+c$$

- Finally substituting back in terms of $x$:
  $= (1/5)(2x-5)^{5/2} + (5/3)(2x-5)^{3/2} + c.$

**Example 10**

Find $\int \frac{x}{(x^2+8)}\,dx$, using the substitution $u = x^2+8$.

**Solution**

To begin, $\frac{du}{dx} = 2x \Rightarrow \frac{du}{2} = x\,dx$

- So: $\int \frac{1\times x}{x^2+8}dx = \int \frac{1}{u}\cdot\frac{du}{2} = \tfrac{1}{2}\int \frac{1}{u}du = \tfrac{1}{2}\ln|u|+c$
- Substituting back in terms of $x := \tfrac{1}{2}\ln|x^2+8|+c.$

Sometimes the **choice of substitution** is left to the student. The substitution should make the integral look easier. Here are two rules of thumb:

- Let $u$ = the expression in the brackets. (Examples 6 and 9)
- If one part of the integral looks like the derivative of another part, then let $u$ = the bit that you can differentiate. (Examples 7, 8 and 10.)

**9** Always be on the look out for **logarithmic integrals**. They are of the form:

$$\int \frac{f'(x)}{f(x)}dx = \ln|f(x)|+c$$

The generalised formula says that if you see an integral which is a rational fraction, where the top is the differential of the bottom, then it is a logarithmic integral.

**Example 11** $\int \frac{8x-5}{4x^2-5x+1}dx$

**Solution** The answer is $\ln|4x^2-5x+1|+c$

because $\frac{d}{dx}(4x^2-5x+1) = 8x-5$, which is on the top of the integral.

Logarithmic integration also works when the top is a multiple of the differential of the bottom. This is shown in the following example:

**Example 12**

Find $\int \frac{x+1}{2x^2+4x+7}\,dx$.

**Solution**

We note that $\frac{d}{dx}(2x^2 + 4x + 7) = 4x + 4$, which is 4 times the numerator.

We proceed by making the numerator what we want it to be; multiplying the integral by ¼ to make the whole thing work. Hence the numerator now becomes the differential of the denominator.

- $\text{So}: \frac{1}{4}\int \frac{4x+4}{2x^2+4x+7}\,dx = \frac{1}{4}\ln\left|2x^2+4x+7\right| + c.$

As an exercise, attempt Example 10 again, using logarithmic integration.

Logarithmic integration also applies to trigonometric expressions:

**Example 13** Prove $\int \tan x \, dx = \ln|\sec x| + c$.

**Solution**

$\int \tan x\, dx = \int \frac{\sin x}{\cos x}\,dx$ and $\int \frac{d}{dx}(\cos x) = -\sin x$.

- So the integral becomes:

$$-\int \frac{-\sin x}{\cos x}\,dx = -\ln\left|\cos x\right| + c = \ln\left|(\cos x)^{-1}\right| + c = \ln\left|\frac{1}{\cos x}\right| + c = \ln\left|\sec x\right| + c.$$

using the laws of logarithms and the fact $\sec x = 1/\cos x$

**10** **Integration by parts** is used when we are asked to integrate a product of two functions which bear no relation to each other, i.e. one is not a differential of the other. We use the formula:

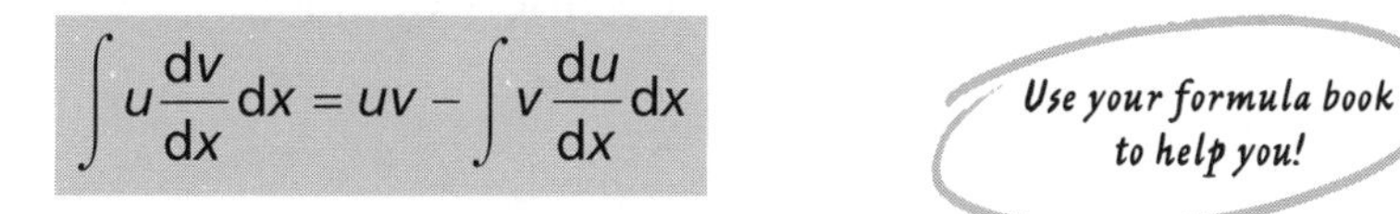


$$\int u\frac{dv}{dx}\,dx = uv - \int v\frac{du}{dx}\,dx$$

A common problem is which function we call $u$ (the function to be differentiated), and which we call $dv/dx$ (to be integrated). This can be resolved by the following rule of thumb:

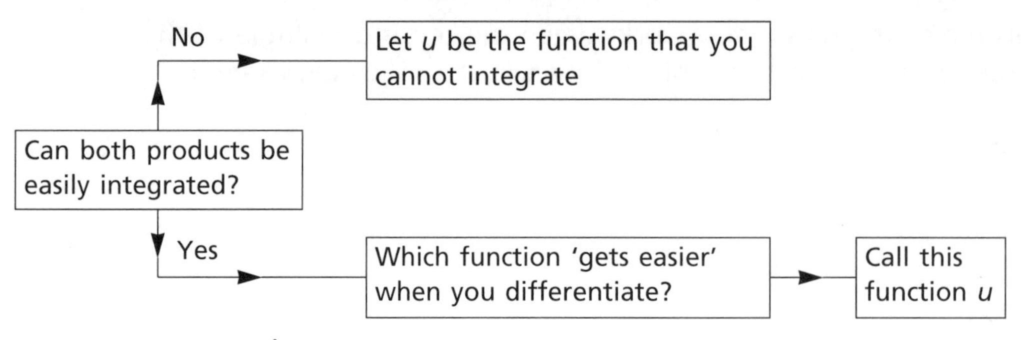

**Example 14** Find $\int xe^{2x}\,dx$.

**Solution**

We have the products $x$ and $e^{2x}$. Both products can easily be integrated, but $x$ becomes easier when you differentiate it (i.e. it becomes 1). Hence:

- Name the functions: $u = x, \quad \dfrac{dv}{dx} = e^{2x}$
- Differentiate and integrate: $\dfrac{du}{dx} = 1, \quad v = \tfrac{1}{2} e^{2x}$
- Apply the formula: $\int x\, e^{2x}\,dx = \tfrac{1}{2} xe^{2x} - \int \tfrac{1}{2} e^{2x}\,dx$
- Integrate: $= \tfrac{1}{2} xe^{2x} - \tfrac{1}{4} e^{2x} + c.$

**Example 15** Find $\int x \ln x\,dx$

**Solution**

We have the products $x$ and $\ln x$. The product $\ln x$ cannot easily be integrated, so it must be called $u$. Hence:

- Differentiate and integrate: $u = \ln x, \quad \dfrac{dv}{dx} = x \quad \Rightarrow \quad \dfrac{du}{dx} = \dfrac{1}{x}, \quad v = \tfrac{1}{2} x^2$
- Apply the formula: $\int x \ln x\,dx = \tfrac{1}{2} x^2 \ln x - \int \left(\dfrac{1}{x}\right) \tfrac{1}{2} x^2\,dx = \tfrac{1}{2} x^2 \ln x - \tfrac{1}{2} \int x\,dx$
- Integrate: $= \tfrac{1}{2} x^2 \ln x - \tfrac{1}{4} x^2 + c.$

Some questions may require the repeated use of integration by parts.

**Example 16** Find $\int x^2 \sin x\,dx$

**Solution**

We have the products $x^2$ and $\sin x$. Since we can integrate both and $x^2$ is the easier of the two to differentiate we let $u = x^2$.

- Hence: $u = x^2, \quad \dfrac{dv}{dx} = \sin x \quad \Rightarrow \quad \dfrac{du}{dx} = 2x, \quad v = -\cos x$
- So: $\int x^2 \sin x \, dx = -x^2 \cos x - \int -2x \cos x \, dx = -x^2 \cos x + \int \mathbf{2x \cos x \, dx}$. ③

It is not possible simply to integrate the part in **bold**. So we must use integration by parts again to evaluate the new problem, $\int 2x \cos x \, dx$.

- So: $u = 2x, \quad \dfrac{dv}{dx} = \cos x \quad \Rightarrow \quad \dfrac{du}{dx} = 2, \quad v = -\sin x$
- Apply the formula:

  $\int 2x \cos x \, dx = 2x \sin x - \int 2\sin x \, dx = 2x \sin x + 2\cos x + c.$ ④
- Substituting ④ into ③: $\int x^2 \sin x \, dx = -x^2 \cos x + 2x \sin x + 2\cos x + c.$

Now learn how to use your knowledge

# Integration

## Use your knowledge

Hints

**1** a) Find $\int \frac{1}{2-x}\,dx$.

*A ln integral*
*Watch the minus*

b) The gradient of a curve is given by the equation $\frac{dy}{dx} = 4 - \frac{1}{x^3} + \frac{1}{2-x}$.
The curve passes through the point (1, 5/2).
Calculate the equation of the curve.

*Integrate to find 'y = …'*
*Use the co-ordinates to find the constant of integration*

**2** a) Given that $f(x) = \frac{8x+7}{(2x^2+7)(4-x)}$ can be expressed in the form $\frac{Ax+B}{2x^2+7} + \frac{C}{4-x}$ find the values of the constants $A$, $B$ and $C$.

*This is a partial fraction question*
*Look at Algebra 2 for help!*

b) Hence evaluate $\int_1^3 f(x)dx$, leaving your answers in the form $\ln D$, where $D$ is an integer.

*Both terms should be ln integrals*
*Remember $\ln 1 = 0$*

**3** a) Find $\int x\cos 3x\,dx$.

*Which product differentiates easily?*

b) Using the substitution, $x = 3\sin u$ rewrite $\int_0^{3/2} \sqrt{(9-x^2)}\,dx$ in the form $k\int_0^a \cos^2 u\,du$, where $k$ and $a$ are constants to be found.

*Use the identity: $\sin^2 x + \cos^2 x = 1$*

*Remember to change the limits*

c) Hence evaluate $\int_0^{3/2} \sqrt{(9-x^2)}\,dx$, leaving your answer in the form $e\sqrt{3} + f\pi$, where $e$ and $f$ are rational constants which need to be found.

*Use the double-angle formula for cosine, on the 'u' version of the integral*

**4** a) Use the method of integration by parts to find $\int we^{-w}\,dw$

*Choose $u = w$.*

b) Using the substitution $x = e^w$, show that

$$\int \frac{3-\ln x}{x^2}\,dx \text{ becomes } \int \frac{3-w}{e^w}\,dw.$$

*Use $x^2 = e^w e^w$, to help you!*

c) Hence find $\int \frac{3-\ln x}{x^2}\,dx$.

*Use the w integral*
*Split up the terms*

Answers on page 91

# Applications of integration

## Test your knowledge

**1** The region $R$ is bounded by the curve $y = x^2 + 6$, the $x$-axis, the $y$-axis, and the line, $n$, with equation $y = -x + 12$.

a) Find the co-ordinates where the curve intersects the line $n$.

b) Draw a sketch, shading in the region $R$ required.

c) Hence calculate the area of the region $R$.

**2** Find the area between the curve $y = 2x^2 - 5x$ and the $x$-axis.

**3** i) Find the particular solution of the differential equation

$$x\frac{dy}{dx} = y^2$$

given that when $x = 1$, $y = 1$.

ii) Find the general solution of the differential equation,

$$\frac{dy}{dx} = e^{2x-y}$$

leaving your answer in the form of exponentials.

iii) Find the general solution of the differential equation,

$$(1 + x^3)\frac{dy}{dx} = x^2y$$

in the form $y = f(x)$.

## Answers

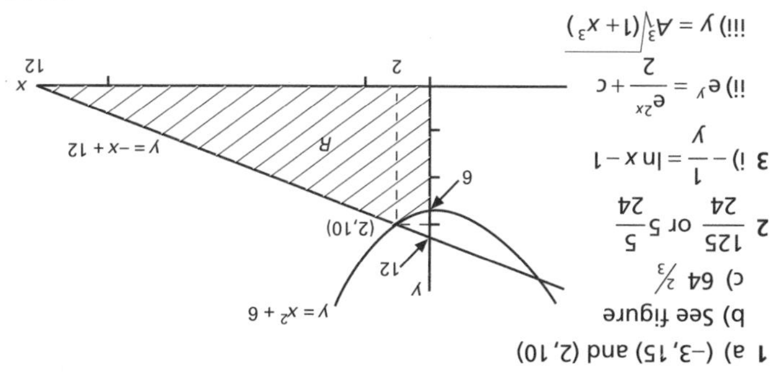


1 a) $(-3, 15)$ and $(2, 10)$

b) See figure

c) $64\tfrac{2}{3}$

2 $\frac{125}{24}$ or $5\frac{5}{24}$

3 i) $-\frac{1}{y} = \ln x - 1$

ii) $e^y = \frac{e^{2x}}{2} + c$

iii) $y = A\sqrt[3]{(1 + x^3)}$

If you got them all right, skip to page 78

# Applications of integration

## Improve your knowledge

**1** One application of integration is to **find area** between **curves** and the co-ordinate **axes**.

**Example 1** $\int_1^4 x^2 dx = \left[\frac{x^3}{3}\right]_1^4 = \left(\frac{64}{3} - \frac{1}{3}\right) = 21 \text{ (units)}^2$

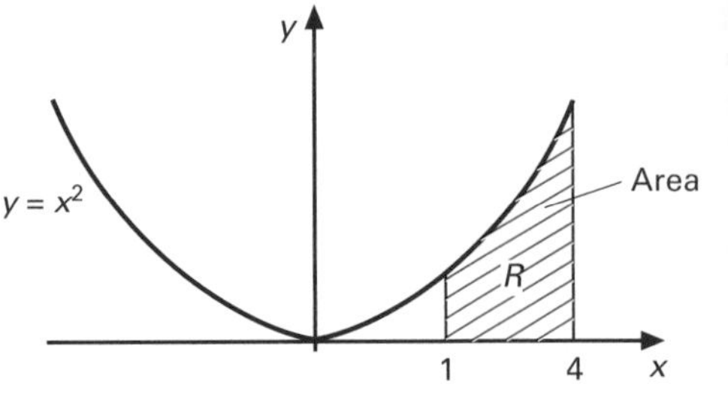


represents the area under the curve $y = x^2$, between the lines $x = 1$ and $x = 4$, as shown in the diagram.

Hence definite integration with respect to $x$ can represent the area between the curve and the $x$-axis.

**Example 2**

The region $R$ is bounded by the curve $y = x^2 + 4$ and the line $l$ with equation $y = -2x + 12$.

a) Find the co-ordinates where the curve intersects the line $l$.

b) Hence calculate the area of the region $R$.

**Solution**

a) Curve intersects $\Rightarrow$ curve = line $\Rightarrow$ solve equations simultaneously.

- Hence: $x^2 + 4 = -2x + 12 \Rightarrow x^2 + 2x - 8 = 0$
  $\Rightarrow (x + 4)(x - 2) = 0 \Rightarrow x = 2$ or $x = -4$
- When $x = 2$, $y = 2^2 + 4 = 8$ and when $x = -4$, $y = (-4)^2 + 4 = 20$
- So the co-ordinates of intersection are: (2, 8) and (–4, 20).

b) To help us to find the area of $R$, we usually draw a sketch to see what is happening.

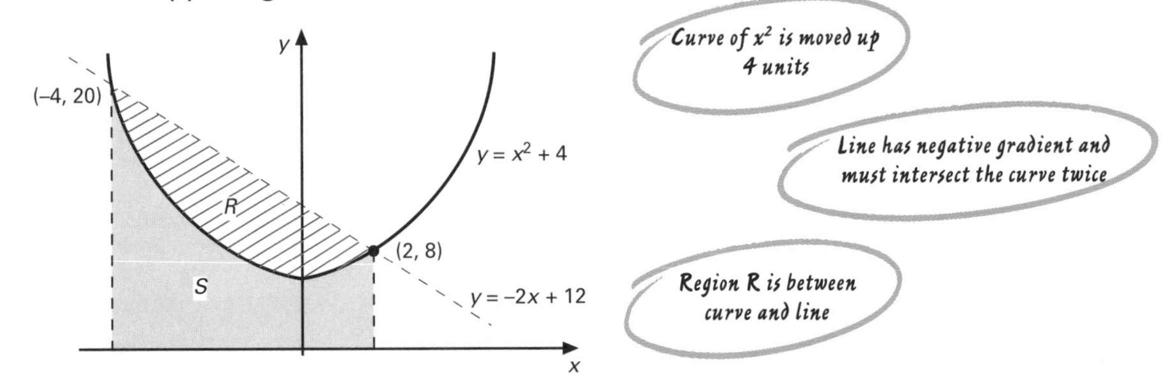

The region $S$ has been shaded in to help us.

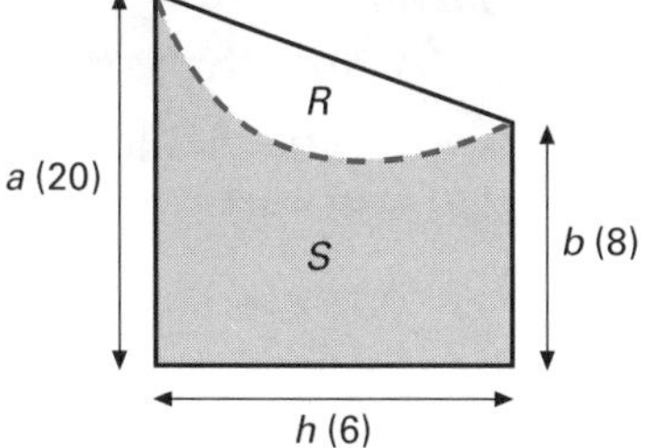


The regions $R$ and $S$ together form a quadrilateral known as a trapezium

- Area of trapezium = ½$(a + b)h$
- Area$(R + S)$ = ½(20 + 8)(6) = 84 (units)$^2$

If we subtract region $S$ from the trapezium, we are left with region $R$. $S$ is found by integrating the curve $y = x^2 + 4$ between $x = -4$ and $x = 2$.

- $\int_{-4}^{2}(x^2+4)\mathrm{d}x = \left[\frac{x^3}{3}+4x\right]_{-4}^{2} = \left(\frac{8}{3}+8\right)-\left(\frac{-64}{3}-16\right) = 48\,\text{units}^2$
- Hence: area $(R)$ = area$(R + S)$ – area$(S)$ = 84 – 48 = 36 units$^2$.

**2** **Integration** will give us a **negative value** if we integrate a region **below the *x*-axis.**

**Example 3** Find the area between the curve $y = x^2 - 4x$ and the $x$-axis.

**Solution**
$y = x(x - 4) = 0 \Rightarrow x = 0$ and $x = 4$, where the curve cuts the x-axis.
It's a happy curve!

- Area required $= \int_{0}^{4}(x^2-4x)\,\mathrm{d}x$

$$= \left[\frac{x^3}{3}-2x^2\right]_0^4 = \left(\frac{64}{3}-32\right) = -\frac{32}{3}$$

$$= -10\frac{2}{3} \text{ which is negative!}$$

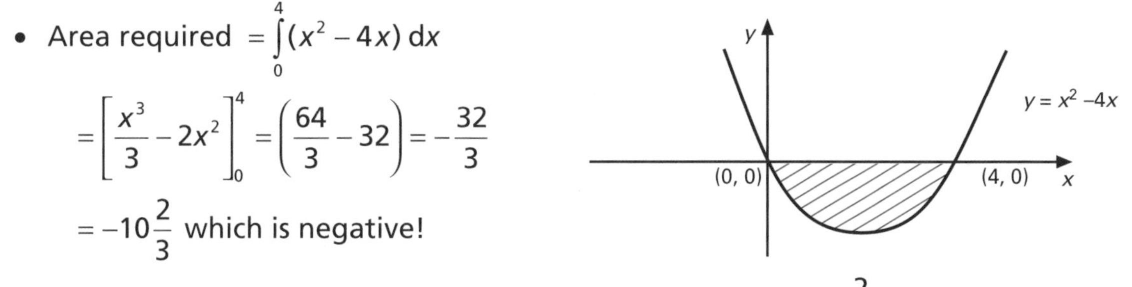


- Since areas must be positive we take the modulus: area $= 10\frac{2}{3}$.

**Example 4**
The diagram shows a sketch of the graph $y = x\ln x$ for $x \geq 0$, which has a minimum stationary point at $A$ and cuts the $x$-axis at $B(1, 0)$.

a) Calculate the co-ordinates of $A$.
b) Find the area enclosed between the line segment $AB$ and the curve.

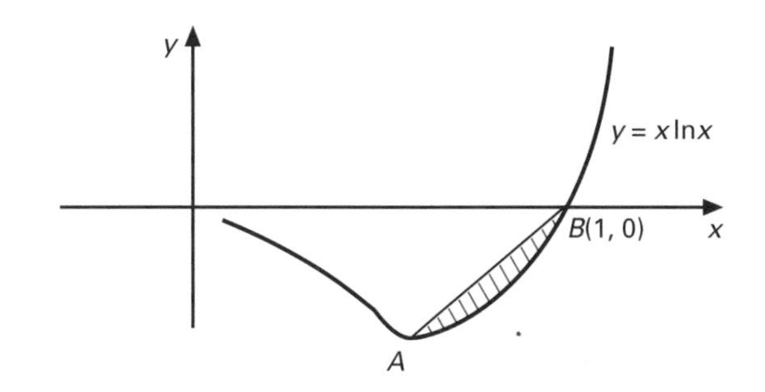

**Solution**

a) To find the stationary point, we differentiate and set to zero.

- Product rule $\Rightarrow u = x \quad v = \ln x$

$$\frac{du}{dx} = 1 \quad \frac{dv}{dx} = \frac{1}{x} \quad \Rightarrow \frac{dy}{dx} = \ln x + 1 = 0$$

$$\Rightarrow \ln x = -1 \Rightarrow x = e^{-1}$$

- Using the $y$-formula: $y = e^{-1}\ln e^{-1} = e^{-1}(-1) = -e^{-1} \Rightarrow A(e^{-1}, -e^{-1})$

b) Region $(R + S)$ is found by integrating the curve from $x = e^{-1}$ to $x = 1$

- Region $(R + S) = \int_{e^{-1}}^{1} x \ln x dx$
- Using integration by parts:

$$u = \ln x, \quad \frac{dv}{dx} = x, \quad \Rightarrow \quad \frac{du}{dx} = \frac{1}{x}, \quad v = \frac{x^2}{2}$$

and $\int x \ln x dx = \frac{x^2}{2}\ln x - \int \frac{x^2}{2}\frac{1}{x}dx = \frac{x^2}{2}\ln x - \int \frac{x}{2}dx = \frac{x^2}{2}\ln x - \frac{x^2}{4}$

- Inserting the limits, region $(R + S) = \left[\frac{x^2}{2}\ln x - \frac{x^2}{4}\right]_{e^{-1}}^{1}$

$= (\tfrac{1}{2}\ln 1 - \tfrac{1}{4}) - (\tfrac{1}{2}e^{-2}\ln e^{-1} - \tfrac{1}{4}e^{-2}) = -\tfrac{1}{4} - (-\tfrac{1}{2}e^{-2} - \tfrac{1}{4}e^{-2}) = \tfrac{3}{4}e^{-2} - \tfrac{1}{4} < 0$

- The value calculated is negative because the region is below the $x$-axis. Since area must be positive, then area$(R + S) = -1.\ (\tfrac{3}{4}e^{-2} - \tfrac{1}{4}) = \tfrac{1}{4} - \tfrac{3}{4}e^{-2}$

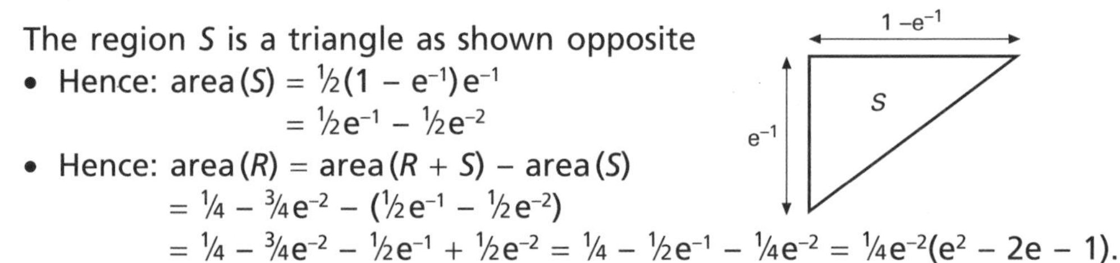


The region $S$ is a triangle as shown opposite

- Hence: area$(S) = \tfrac{1}{2}(1 - e^{-1})e^{-1}$
$\qquad = \tfrac{1}{2}e^{-1} - \tfrac{1}{2}e^{-2}$
- Hence: area$(R)$ = area$(R + S)$ − area$(S)$
$\qquad = \tfrac{1}{4} - \tfrac{3}{4}e^{-2} - (\tfrac{1}{2}e^{-1} - \tfrac{1}{2}e^{-2})$
$\qquad = \tfrac{1}{4} - \tfrac{3}{4}e^{-2} - \tfrac{1}{2}e^{-1} + \tfrac{1}{2}e^{-2} = \tfrac{1}{4} - \tfrac{1}{2}e^{-1} - \tfrac{1}{4}e^{-2} = \tfrac{1}{4}e^{-2}(e^2 - 2e - 1).$

**3** **Differential equations** (DEs) are usually written in the form

$$\frac{dy}{dx} = \text{(something in terms of } y \text{ and } x\text{)}.$$

To solve a DE, you must integrate to find a relationship between $y$ and $x$, with $\frac{dy}{dx}$ disappearing!

In the exam you only need to know the method of separation. This means getting all the $y$ terms on one side with d$y$ on the top, and getting all the $x$ terms on the other side with d$x$ on the top. Then you can integrate both sides, one with respect to $y$ and the other with respect to $x$.

**Example 5**

Find the general of the DE, $(1 + x^2)\dfrac{dy}{dx} = xy$ in the form $y = f(x)$. Hence find the particular solution when $x = 0$, $y = 5$ to the differential equation.

**Solution**

The **general solution** is the solution to the DE expressed with an unknown constant.

- Separate the $x$'s and the $y$'s: $\dfrac{1}{y}dy = \dfrac{x}{1+x^2}dx$
- Put an integral sign on both sides: $\displaystyle\int \frac{1}{y}dy = \int \frac{x}{1+x^2}dx$
- Fix the RHS: $\displaystyle\int \frac{1}{y}dy = \frac{1}{2}\int \frac{2x}{1+x^2}dx$

The RHS is a logarithmic integral because $\dfrac{d}{dx}(1+x^2) = 2x$

- Integrate both sides: $\ln y = \frac{1}{2}\ln(1 + x^2) + c$

Before we can take e (or exponentiate) both sides we say $c = \ln A$ where $A$ is another constant, so $\ln A$ is another constant.

- So: $\ln y = \ln(1 + x^2)^{1/2} + \ln A \Rightarrow \ln y = \ln(A(1 + x^2)^{1/2})$ by the addition law of logarithms
- Exponentiating both sides: $y = A(1 + x^2)^{1/2} = A\sqrt{(1+x^2)}$

This is the general solution. The **particular solution** is found when you solve the DE and find the constant, say $A$.

- Use the initial conditions, $x = 0$, $y = 5$, to find the constant $A$: $5 = A\sqrt{(1+0^2)} \Rightarrow A = 5$
- State the particular solution: $y = 5\sqrt{(1+x^2)}$.

**Example 6**

Find the general solution of the DE $\dfrac{dy}{dx} = e^{x+y}$.

**Solution**

This looks a nightmare because it seems impossible to split the $x$ and $y$ terms. But, by the power of indices, we say $e^{x+y} = e^x e^y$, giving $\dfrac{dy}{dx} = e^x e^y$.

- Separating the variables: $\displaystyle\int \frac{1}{e^y}dy = \int e^x dx \Rightarrow \int e^{-y}dy = \int e^x dx$ since $\dfrac{1}{e^y} = e^{-y}$
- Hence the general solution is: $-e^{-y} = e^x + c$.

*Now learn how to use your knowledge*

# Applications of integration

35 minutes

## Use your knowledge

Hints

**1** The diagram below shows the graph of $y = \frac{(x-1)(x-5)}{(x+2)}$, which cuts the $x$-axis at the points $W$ and $X$. The points $Y$ and $Z$ are located on the curve when $x = 2$ and $x = 4$ respectively. The region $R$ shown in the figure is bounded by the curve $x = 2$, $x = 4$ and the line segment $YZ$.

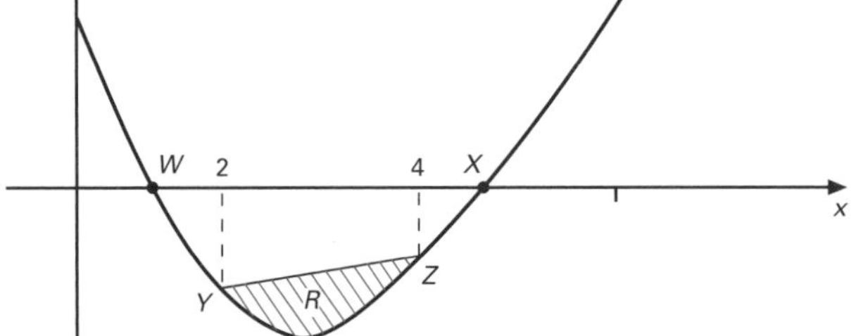


a) Write down the co-ordinates of $W$, $X$, $Y$ and $Z$.

*Set $y = 0$ to find $W$ and $X$*

b) Express the equation $y = \frac{(x-1)(x-5)}{(x+2)}$ in the form:

$$y = (Ax + B) + \frac{C}{(x+2)}$$

where $A$, $B$ and $C$ are constants to be determined.

*Treat this as a partial fractions question*

c) Hence, or otherwise, show that the area of the region $R$ is

$$\frac{35}{4} - 21\ln\left(\frac{3}{2}\right).$$

*Integrate part b) between $x = 2$ and $x = 4$. Then subtract out the trapezium*

**2** a) Show that $\int x^2 e^{-3x} dx$ can be expressed in the form

$$-\frac{1}{27} e^{-3x} (f(x))$$

where $f(x)$ is to be determined.

*Use integration by parts twice!*

b) Find the particular solution in terms of exponentials of the differential equation $\frac{dy}{dx} = x^2 e^{y-3x}$, given that $x = 0$, $y = 0$.

*Look at part a) for help! Find constant by using $x = 0$, $y = 0$.*

Answers on page 92

# Numerical methods

## Test your knowledge

**1** a) Show that the equation $x - \sin x = {}^{2\pi}\!/_3$ has a root in the interval [2, 3] and find the root correct to 1 decimal place.

b) The equation $x^4 - 13x^2 - x + 30 = 0$ has two roots between 0 and 5. Find an interval of length 1 containing each root.

**2** a) By drawing suitable sketch graphs, find the number of solutions to the equations:

i) $x = 2\sin 2x$ ii) $e^x + \ln x = 0$.

**3** a) Use the iteration formula $x_{n+1} = 0.5\cos(x_n)$ and $x_1 = 1$ to find the solution to the equation $2x = \cos x$, giving your answer correct to 2 decimal places.

b) i) Show that the equation $x^3 + 4x^2 - 1 = 0$ can be rearranged into the form $x = \sqrt{\dfrac{A}{x+B}}$ where $A$ and $B$ are constants to be found.

ii) Using $x_1 = 1$, use an iteration formula derived from your answer to b) i) to find an approximate solution to the equation $x^3 + 4x^2 - 1 = 0$, giving your answer correct to 3 decimal places.

### Answers

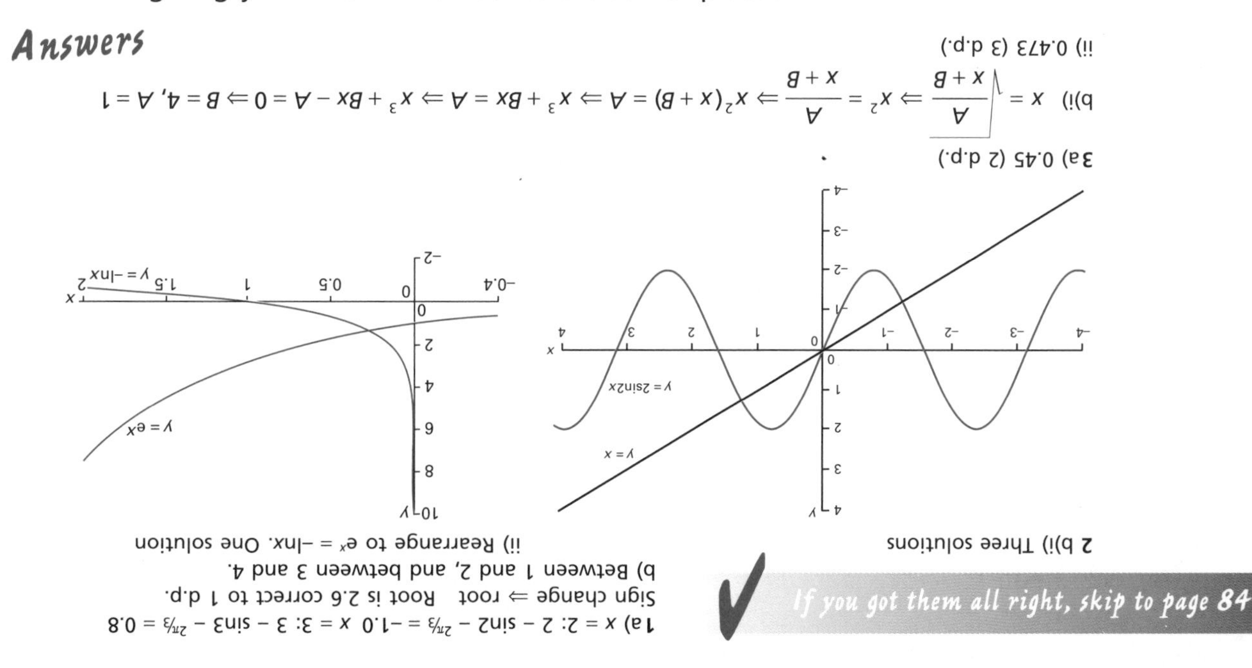


**1** a) $x = 2$: $2 - \sin 2 - {}^{2\pi}\!/_3 = -1.0$ $x = 3$: $3 - \sin 3 - {}^{2\pi}\!/_3 = 0.8$
Sign change $\Rightarrow$ root Root is 2.6 correct to 1 d.p.
b) Between 1 and 2, and between 3 and 4.

**2** b)i) Three solutions

ii) Rearrange to $e^x = -\ln x$. One solution

**3** a) 0.45 (2 d.p.)

b)i) $x = \sqrt{\dfrac{A}{x+B}} \Rightarrow x^2 = \dfrac{A}{x+B} \Rightarrow x^2(x+B) = A \Rightarrow x^3 + Bx = A \Rightarrow x^3 + Bx - A = 0 \Rightarrow B = 4, A = 1$

ii) 0.473 (3 d.p.)

If you got them all right, skip to page 84

# Numerical methods

Numerical methods are used to find the solution to equations that you can't solve accurately. They only ever give approximate solutions. (**Exam questions will always make it clear when they want you to solve equations numerically**.) Anything to do with **trigonometry MUST** be in **radians** when you use numerical methods.

Three methods will be covered here:

**1** The **sign change** method is for finding a solution of an equation of the form $f(x) = 0$, where $f(x)$ can be any expression at all containing $x$.

- Any other equation needs to be rearranged into this form first. So $x = \tan x$ would go to $x - \tan x = 0$.
- It tells you the root lies in a particular interval – say between 2 and 3.
- The best estimate for the root is the midpoint of the interval.
- If you are trying to show that a root lies between 2 and 3, you put 2 and 3 in the equation, and write down what answers you get. **If the answers are of different signs, there is a root in the interval.** .
- You can make your answer as accurate as you like by looking at smaller intervals.

*If a question asks you to show a root lies between two numbers, use the sign-change method*

*Some graphic calculators will do a lot of this for you. But you must still show your working to get the marks!*

**Example 1**
The equation $e^x = 5x$ has two roots between 0 and 4.
a) Show that the smaller root lies between 0 and 1.
b) Find an interval of length 1 in which the larger root lies.
c) Find the smaller root correct to 1 decimal place.

**Solution**
Before we can start we must rearrange the equation, so it has 0 on one side. So $e^x - 5x = 0$.

a) We need to show there is a sign change. Try 0: $e^0 - 5(0) = 1$
Try 1: $e^1 - 5(1) = -2.28$ So there is a sign change, so we have a root.

b) We need to find another interval with a sign change, so try 2, 3, 4 in order (we know the root is less than 4):
$e^2 - 5(2) = -2.61$ – not between 1 and 2, since both negative.
$e^3 - 5(3) = 5.09$ – changes from negative to positive between 2 and 3 – so the root is between 2 and 3.

*You must actually write the values down – not just say they are positive or negative*

c) We know the root is between 0 and 1 but need to find smaller intervals, so we try numbers between 0 and 1.
$e^{0.5} - 5(0.5) = -0.85$ – so there is a sign change between 0 and 0.5, so root is between them
$e^{0.25} - 5(0.25) = 0.03$ – so root is between 0.25 and 0.5
$e^{0.4} - 5(0.4) = -0.51$ – so root is between 0.25 and 0.4
$e^{0.3} - 5(0.3) = -0.15$ – the root is between 0.25 and 0.3, so it is 0.3 to 1 decimal place.

**2** **The graphical method** is for finding a solution of an equation of the form $f(x) = g(x)$, where $f(x)$ and $g(x)$ are both expressions involving $x$.

- Both $f(x)$ and $g(x)$ should be functions whose graphs are relatively easy to draw. So $5x = \sin x$ is OK, but $\sin x = x\tan x$ is not, because $x\tan x$ is not easy to draw.
- The root is the $x$-value of the crossing point of the graphs of the two sides of the equation.

Graphical methods are often used to find **how many** roots an equation has, rather than finding them accurately.

*Graphic calculators can be very useful here – but if a question specifies a graphical method, there will be no marks for the answer alone*

**Example 2**

By sketching appropriate graphs, find the number of roots of the equations

a) $\ln x = 2 - x$ b) $x^2 = \sin x$ c) $xe^x = 1$.

**Solution**

a) Since we are asked to **sketch** a graph, we do not have to use graph paper.

From the graph we can see there is 1 root of the equation $\ln x = 2 - x$

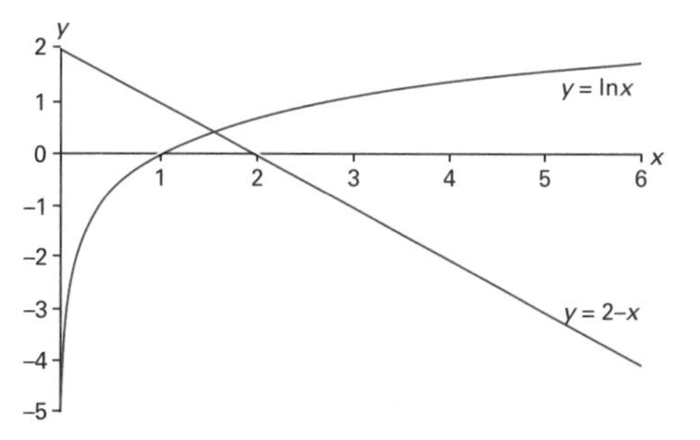

b) We must work in radians for the two graphs.

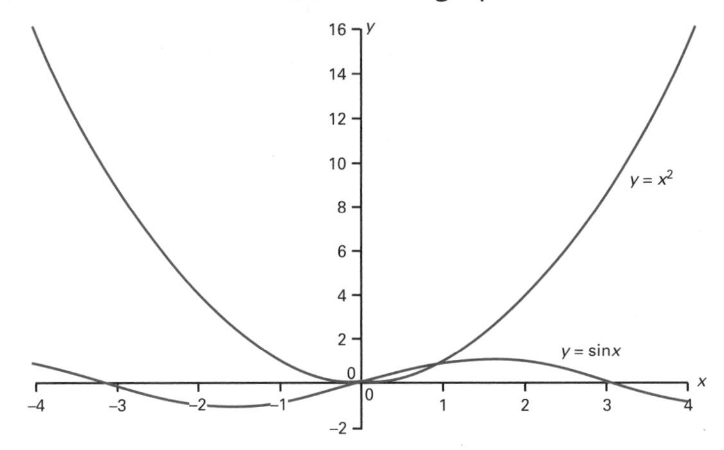


The graph shows there are 2 roots of the equation $x^2 = \sin x$.

c) As it stands, this is not easy to do because $y = xe^x$ does not have a simple graph. We therefore rearrange the equation:
$xe^x = 1 \Rightarrow e^x = 1/x$ (or, if preferred, $x = e^{-x}$)

d) Draw the graphs of $y = 1/x$ and $y = e^x$

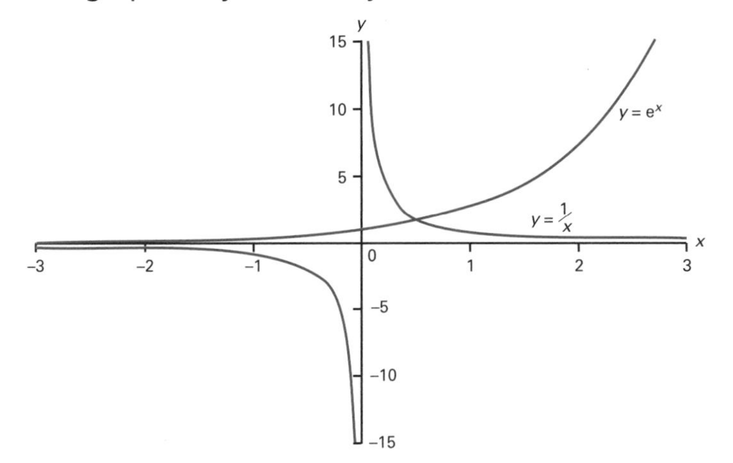


There is 1 root of the equation $e^x = 1/x$ and hence of the equation $xe^x = 1$.

**3** **Iteration** is a method for finding a solution of an equation of the form $x = f(x)$ where $f(x)$ is any expression involving $x$.

*Again – check whether your graphic calculator can do this!*

- Any other equations must be rearranged into this form first. So $x^2 - 5\sin x = 0$ becomes $x = \sqrt{(5\sin x)}$.
- You have to have a starting value (or 'first guess') for the solution; this is called $x_1$. This will either be given to you in the question, or you'll be told to find it using the sign-change or graphical method.

- You use the formula $x_{n+1} = f(x_n)$, (from the equation you started with), to produce values $x_2, x_3, \ldots$ which should be better and better approximations. So the equation $x = \sqrt{(5\sin x)}$ would give the formula $x_{n+1} = \sqrt{(5\sin x_n)}$.
- The question may tell you when to stop; alternatively, if it requires the answer correct to 3 decimal places, say, you stop when two successive answers agree to 3 decimal places.

*If you need the answer to 3 d.p. all working **MUST** be done correct to at least 4 d.p.*

This process does not always work. If it **works**, it is said to **converge**, if it does **not work** (which you can tell because the values move away from each other) it is said to **diverge**.

**Example 3**

Using the iterative formula $x_{n+1} = e^{-x_n}$ and $x_1 = 0.5$, find, correct to 2 decimal places, the solution to the equation $x = e^{-x}$.

**Solution**

Since the answer needs to be to 2 decimal places, we must work to 3 decimal places. We know that $x_1 = 0.5$.

Using the formula:

$x_2 = e^{-x_1} = e^{-0.5} = 0.607$; $x_3 = e^{-x_2} = e^{-0.607} = 0.545$; $x_4 = e^{-x_3} = e^{-0.545} = 0.580$;
$x_5 = e^{-x_4} = e^{-0.580} = 0.560$; $x_6 = e^{-x_5} = e^{-0.560} = 0.571$; $x_7 = e^{-x_6} = e^{-0.571} = 0.565$;
$x_8 = e^{-x_7} = e^{-0.565} = 0.568$

Since our last two answers agree to 2 decimal places, the solution is 0.57 correct to 2 decimal places.

Questions may also require you to produce an iteration formula by rearranging an equation into the appropriate form. If you can see how to do it, that's fine, but Example 4 shows a 'cheat' method if you can't.

**Example 4**

Show that the equation $x^3 - 5x^2 - 6 = 0$ can be rearranged into the form

$x = \sqrt{\dfrac{Ax^2 + B}{x}}$, where A and B are constants to be determined.

**Solution**

Since it is not at all obvious how to get this, we try working backwards!

$$x = \sqrt{\frac{Ax^2 + B}{x}} \Rightarrow x^2 = \frac{Ax^2 + B}{x} \Rightarrow x^3 = Ax^2 + B \Rightarrow x^3 - Ax^2 - B = 0$$

So: $A = 5$, $B = 6$ from looking at the original equation.

*Now learn how to use your knowledge*

# Numerical methods

## Use your knowledge

**Hints**

**1** a) Show that the equation $x^2 = 2$ can be rearranged into the form $x = A + \frac{B}{x+1}$, where $A$ and $B$ are constants to be found.

*Multiply the equation through by $(x + 1)$ then rearrange*

b) Using $x_1 = 1$, use an iteration formula derived from your answer to a) to find the value of $\sqrt{2}$ correct to 4 decimal places.

*Work to 5 d.p. Stop when your answers agree to 4 d.p.*

**2** a) i) Sketch, on the same diagram, the graphs of $y = \frac{1}{5x}$ and $y = x^2 - 2$.

*Refer to Chapter 3 Functions if you have a problem here!*

ii) Hence state the number of solutions of the equation $5x(x^2 - 2) = 1$.

*Link this with the graphs!*

b) Show that one solution of this equation lies between 0 and –1, and find the solution correct to the nearest 0.2.

*Sign-change method! Is it closest to 0, –0.2, –0.4...*

**3** a) Using an appropriate sketch graph, explain why the equation $\tan x = x$ has infinitely many solutions.

*Sketch $y = x$ and $y = \tan x$ How many times will it cross every 180°?*

b) By using the sign-change method, or otherwise, find the solution to this equation in the interval $[\pi, \frac{3\pi}{2}]$ correct to the nearest 0.2.

*Work in radians!*

c) i) Using your answer to b) for $x_1$, use the iteration formula $x_{n+1} = \tan^{-1} x_n$ to find $x_2$ and $x_3$.

ii) Is this iteration formula suitable for finding the root in the interval $[\pi, \frac{3\pi}{2}]$ accurately? Explain your answer.

*Look at the values you are getting!*

Answers on page 94

## Algebra 1

**1**a) Using formula:

$x = [6 \pm \sqrt{(36-8)}]/2 = [6 \pm \sqrt{28}]/2 = [6 \pm \sqrt{4}\sqrt{7}]/2 = [6 \pm 2\sqrt{7}]/2 = 3 \pm \sqrt{7}$

b) $3-\sqrt{7}$ $\quad$ $3+\sqrt{7}$

Put in $x = 0$: $(0)^2 - 6(0) + 2 = 2$ +ve $\quad x = 3$: $(3)^2 - 6(3) + 2 = -7$ –ve $\quad x = 6$: $(6)^2 - 6(6) + 2 = 2$ +ve

So +ve $\quad 3-\sqrt{7}$ $\quad$ –ve $\quad 3+\sqrt{7}$ $\quad$ +ve

So require $3 - \sqrt{7} \leq x \leq 3 + \sqrt{7}$

**2** Let length = $x$. Then width = $x - 4$.
Perimeter = $2x + 2(x - 4) = 4x - 8 \leq 36$ (**1**) $\quad$ Area = $x(x - 4) \geq 60$ (**2**)
(**1**) $\Rightarrow 4x \leq 44 \Rightarrow x \leq 11$
(**2**) $\Rightarrow x^2 - 4x - 60 \geq 0 \Rightarrow (x - 10)(x + 6) \geq 0$ $\quad$ +ve $\quad -6$ $\quad$ –ve $\quad 10$ $\quad$ +ve

So require $x \geq 10$ (since can't have negative length)
So combining (**1**) and (**2**), require $x \geq 10$ and $x \leq 11$ $\quad$ so $10 \leq x \leq 11$

**3**a) $P = -Q + 2\ln 4$

b) $\ln y = -\ln x + 2\ln 4 \Rightarrow \ln y = \ln 4^2 - \ln x \Rightarrow \ln y = \ln(^{16}/_x) \Rightarrow y = {}^{16}/_x$

**4**a)

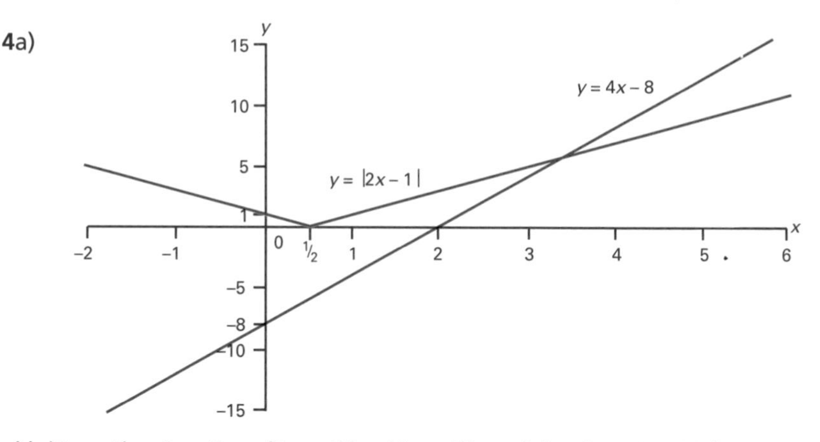


b) $|2x - 1| = 4x - 8 \Rightarrow (2x - 1)^2 = (4x - 8)^2 \Rightarrow 4x^2 - 4x + 1 = 16x^2 - 64x + 64$
$\Rightarrow 12x^2 - 60x + 63 = 0 \Rightarrow (6x - 9)(2x - 7) = 0$
$\Rightarrow x = {}^3/_2$ or ${}^7/_2$. From graph, only $x = {}^7/_2$ is genuine, so this is the solution

c) $y = |2x - 1|$ is above $y = 4x - 8$ for $x < {}^7/_2$

**5**a) $2x^2 + 20x - 15 \equiv A(x + B)^2 + C \equiv A(x^2 + 2Bx + B^2) + C \equiv Ax^2 + 2ABx + AB^2 + C$

So, looking at $x^2$: $\quad 2x^2 \equiv Ax^2 \quad \Rightarrow A = 2$
Looking at $x$: $\quad 20x \equiv 2ABx \quad \Rightarrow 2AB = 20 \Rightarrow 4B = 20 \Rightarrow B = 5$
Constants: $\quad -15 \equiv AB^2 + C \quad \Rightarrow -15 = 50 + C \Rightarrow C = -65$
So $2x^2 + 20x - 15 \equiv 2(x + 5)^2 - 65$

b) We have $2(x + 5)^2 - 65 = 31 \Rightarrow 2(x + 5)^2 = 96 \Rightarrow (x + 5)^2 = 48 \Rightarrow x + 5 = \pm\sqrt{48} = \pm 4\sqrt{3} \Rightarrow x = -5 \pm 4\sqrt{3}$

**6** $2^{x+1} = 2^x \times 2^1 = 2y$. $2^{-x} = \frac{1}{2^x} = \frac{1}{y}$
So $2^{x+1} + 2^{-x} = 3 \Rightarrow 2y + \frac{1}{y} = 3 \Rightarrow 2y^2 + 1 = 3y \Rightarrow 2y^2 - 3y + 1 = 0 \Rightarrow (2y - 1)(y - 1) = 0 \Rightarrow y = \frac{1}{2}, 1$
But $y = 2^x$ $y = \frac{1}{2} \Rightarrow 2^x = \frac{1}{2} \Rightarrow x = -1$ $y = 1 \Rightarrow 2^x = 1 \Rightarrow x = 0$

**7** Let $y = e^x$. Then $e^{2x} = (e^x)^2 = y^2$
So $2e^{2x} - 11e^x + 15 = 0 \Rightarrow 2y^2 - 11y + 15 = 0 \Rightarrow (2y - 5)(y - 3) = 0 \Rightarrow y = \frac{5}{2}, 3$
But $y = e^x$ $y = \frac{5}{2} \Rightarrow e^x = \frac{5}{2} \Rightarrow x = \ln(\frac{5}{2})$ $y = 3 \Rightarrow e^x = 3 \Rightarrow x = \ln 3$

## Algebra 2

**1a)** $(1 - 4x)^7 \equiv 1^7 + {}^7C_1\, 1^6(-4x) + {}^7C_2\, 1^5(-4x)^2 + {}^7C_3\, 1^4(-4x)^3 + \ldots$
$\approx 1 - 28x + 336x^2 - 2240x^3$

b) i) $0.96^7 = (1 - 4x)^7 \Rightarrow 0.96 = 1 - 4x \Rightarrow x = 0.01$
So $(0.96)^7 \approx 1 - 28(0.01) + 336(0.01)^2 - 2240(0.01)^3 = 0.75136$
ii) $0.96^7 = 0.75145$ (5 d.p.) So % error $= (0.75136 - 0.75145) \div 0.75145 \times 100\ \% = -0.012\ \%$

c) First find $(1 + 4x)^7 \equiv 1 + 28x + 336x^2 + 2240x^3 + \ldots$
So $(1 + 4x)^7 + (1 - 4x)^7 \approx 2 + 672x^2$
Putting $x = \frac{1}{\sqrt{3}}$: $(1 + 4(\frac{1}{\sqrt{3}}))^7 + (1 - 4(\frac{1}{\sqrt{3}}))^7 \approx 2 + 672(\frac{1}{\sqrt{3}})^2 = 2 + 224 = 226$

**2** Try $\pm1, \pm2$: $x = 1$: $(1)^3 - (1)^2 - (1) - 2 = -3$ $(x - 1)$ not a factor
$x = -1$: $(-1)^3 - (-1)^2 - (-1) - 2 = -3$ $(x + 1)$ not a factor $x = 2$: $(2)^3 - (2)^2 - (2) - 2 = 0$ $(x - 2)$ is a factor
So $x^3 - x^2 - x - 2 \equiv (x - 2)(x^2 + Ax + 1)$.
So $-x^2 \equiv -2x^2 + Ax^2$ so $A = 1$ $\Rightarrow x^3 - x^2 - x - 2 \equiv (x - 2)(x^2 + x + 1)$.
But $x^2 + x + 1 = 0$ has no real solutions, since $b^2 - 4ac = 1 - 4 < 0$ So $x = 2$ is the only root

**3a)** Put $x = -1$: $-1 + A - B + 6 = 0$ $\Rightarrow A - B = -5$ (1)
Put $x = 2$: $8 + 4A + 2B + 6 = 60$ $\Rightarrow 2A + B = 23$ (2)
(1) + (2) $\Rightarrow 3A = 18 \Rightarrow A = 6$. Substituting in (1): $6 - B = -5 \Rightarrow B = 11$

b) $f(x) \equiv x^3 + 6x^2 + 11x + 6 \equiv (x + 1)(x^2 + Ax + 6)$
$6x^2 \equiv x^2 + Ax^2 \Rightarrow A = 5 \Rightarrow f(x) \equiv x^3 + 6x^2 + 11x + 6 \equiv (x + 1)(x^2 + 5x + 6) \equiv (x + 1)(x + 2)(x + 3)$

**4a)** $x = -1$ is a solution if the answer is zero when $x = -1$ is substituted:
$2(-1)^3 - 7(-1)^2 - 5(-1) + 4 = -2 - 7 + 5 + 4 = 0$
So $2x^3 - 7x^2 - 5x + 4 \equiv (x + 1)(2x^2 + Ax + 4)$
So $-7x^2 \equiv 2x^2 + Ax^2$. So $A = -9$
$\Rightarrow 2x^3 - 7x^2 - 5x + 4 \equiv (x + 1)(2x^2 - 9x + 4) \equiv (x + 1)(2x - 1)(x - 4)$
So other solutions are $x = \frac{1}{2}$ and $x = 4$

b) . −ve **−1** +ve **0.5** −ve **4** +ve .
So solution is $-1 \le x \le 0.5$ and $x \ge 4$

**5a)** $\dfrac{2x - 5}{x - 4} \equiv A + \dfrac{B}{x - 4} \equiv \dfrac{A}{1} + \dfrac{B}{x - 4} \equiv \dfrac{A(x - 4) + B}{x - 4} \Rightarrow 2x - 5 \equiv A(x - 4) + B$

$x = 4$: $8 - 5 = A(0) + B \Rightarrow B = 3$

$x = 0$: $-5 = -4A + B$ But $B = 3 \Rightarrow -5 = -4A + 3 \Rightarrow A = 2$

b) $\dfrac{-3x}{2(x - 1)(x - 4)} \equiv \dfrac{A}{x - 1} + \dfrac{B}{x - 4} \equiv \dfrac{A(2(x - 4)) + B(2(x - 1))}{2(x - 1)(x - 4)} \Rightarrow -3x \equiv 2A(x - 4) + 2B(x - 1)$

$x = 4$: $-12 = 2A(4 - 4) + 2B(4 - 1) \Rightarrow -12 = 6B \Rightarrow B = -2$

$x = 1$: $-3 = 2A(1 - 4) + 2B(1 - 1) \Rightarrow -3 = -6A \Rightarrow A = \frac{1}{2}$

## Functions

**1**a) See figure below
Asymptotes: $y = a$, $x = -b$

b) See figure below

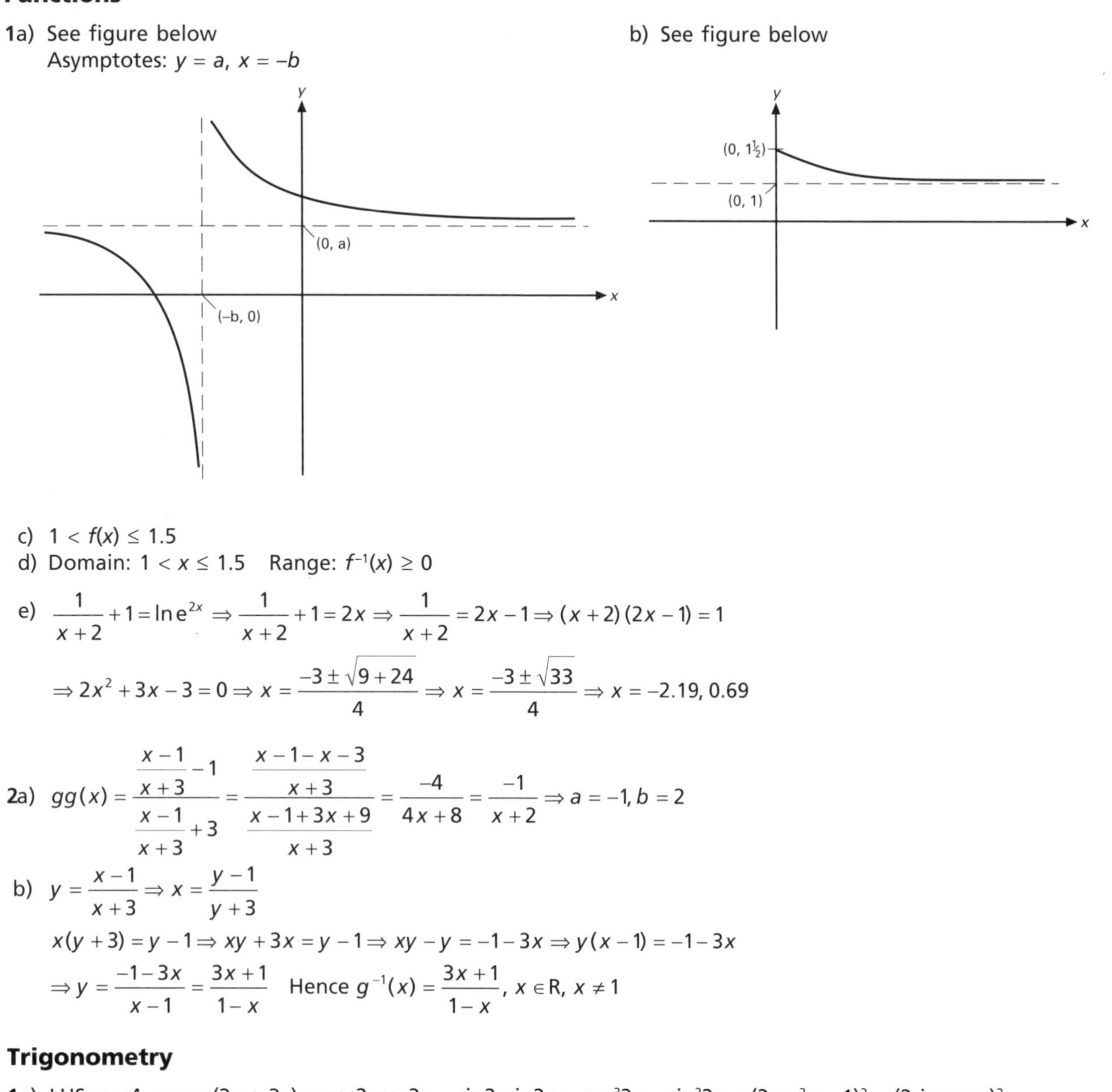


c) $1 < f(x) \leq 1.5$

d) Domain: $1 < x \leq 1.5$ Range: $f^{-1}(x) \geq 0$

e) $\frac{1}{x+2} + 1 = \ln e^{2x} \Rightarrow \frac{1}{x+2} + 1 = 2x \Rightarrow \frac{1}{x+2} = 2x - 1 \Rightarrow (x+2)(2x-1) = 1$

$\Rightarrow 2x^2 + 3x - 3 = 0 \Rightarrow x = \frac{-3 \pm \sqrt{9+24}}{4} \Rightarrow x = \frac{-3 \pm \sqrt{33}}{4} \Rightarrow x = -2.19, 0.69$

**2**a) $gg(x) = \frac{\frac{x-1}{x+3} - 1}{\frac{x-1}{x+3} + 3} = \frac{\frac{x-1-x-3}{x+3}}{\frac{x-1+3x+9}{x+3}} = \frac{-4}{4x+8} = \frac{-1}{x+2} \Rightarrow a = -1, b = 2$

b) $y = \frac{x-1}{x+3} \Rightarrow x = \frac{y-1}{y+3}$

$x(y+3) = y - 1 \Rightarrow xy + 3x = y - 1 \Rightarrow xy - y = -1 - 3x \Rightarrow y(x-1) = -1 - 3x$

$\Rightarrow y = \frac{-1-3x}{x-1} = \frac{3x+1}{1-x}$ Hence $g^{-1}(x) = \frac{3x+1}{1-x}$, $x \in \mathrm{R}$, $x \neq 1$

## Trigonometry

**1**a) LHS: $\cos 4x \equiv \cos(2x + 2x) \equiv \cos 2x \cos 2x - \sin 2x \sin 2x \equiv \cos^2 2x - \sin^2 2x \equiv (2\cos^2 x - 1)^2 - (2\sin x \cos x)^2$
$\equiv 4\cos^4 x - 4\cos^2 x + 1 - 4\sin^2 x \cos^2 x \equiv 4\cos^4 x - 4\cos^2 x + 1 - 4(1 - \cos^2 x)\cos^2 x$
$\equiv 4\cos^4 x - 4\cos^2 x + 1 - 4\cos^2 x + 4\cos^4 x \equiv 8\cos^4 x - 8\cos^2 x + 1 \equiv$ RHS

b) $\cos 4x = \cos^2 x \Rightarrow 8\cos^4 x - 8\cos^2 x + 1 = \cos^2 x \Rightarrow 8\cos^4 x - 9\cos^2 x + 1 = 0$
$\Rightarrow (8\cos^2 x - 1)(\cos^2 x - 1) = 0 \Rightarrow \cos^2 x = \frac{1}{8}$ or $1$
$\cos^2 x = \frac{1}{8} \Rightarrow \cos x = \pm\frac{1}{\sqrt{8}} \Rightarrow x = \pm 69.3°, \pm 110.7°, \pm 249.3°, \pm 290.7°$
$\cos^2 x = 1 \Rightarrow \cos x = \pm 1 \Rightarrow x = 0°, \pm 180°, \pm 360°$

c) Let $x = 2A$. Then $\cos 2x = \cos^2(\frac{1}{2}x)$; $-720 \leq x \leq 720° \Rightarrow \cos 4A = \cos^2(A)$; $-360° \leq x \leq 360°$
So $A = 0°, \pm 69.3°, \pm 110.7°, \pm 180°, \pm 249.3°, \pm 290.7°, \pm 360°$ (from part b).
So $x = 0°, \pm 138.6°, \pm 221.4°, \pm 360°, \pm 498.6°, \pm 581.4°, \pm 720°$.

**2** $2\cos x \cos \frac{\pi}{3} = 2\sin x \sin \frac{\pi}{3} + 1 \Rightarrow 2\cos x \cos \frac{\pi}{3} - 2\sin x \sin \frac{\pi}{3} = 1 \Rightarrow \cos x \cos \frac{\pi}{3} - \sin x \sin \frac{\pi}{3} = \frac{1}{2} \Rightarrow \cos(x + \frac{\pi}{3}) = \frac{1}{2}$
So $x + \frac{\pi}{3} = \frac{\pi}{3}, -\frac{\pi}{3} \Rightarrow x = 0, -\frac{2\pi}{3}$

3a) i) $\tan 2A = \dfrac{2\tan A}{1-\tan^2 A}$ so setting A = ½x: $\tan x = \dfrac{2\tan\frac{x}{2}}{1-\tan^2\frac{x}{2}} = \dfrac{2t}{1-t^2}$

ii) $\sec^2 x \equiv 1+\tan^2 x \Rightarrow \sec^2 x \equiv 1+\dfrac{4t^2}{(1-t^2)^2} \equiv \dfrac{(1-t^2)^2+4t^2}{(1-t^2)^2} \equiv \dfrac{1-2t^2+t^4+4t^2}{(1-t^2)^2} \equiv \dfrac{1+2t^2+t^4}{(1-t^2)^2} = \dfrac{(1+t^2)^2}{(1-t^2)^2}$

Hence $\sec x \equiv \sqrt{\sec^2 x} \equiv \dfrac{(1+t^2)}{(1-t^2)}$

b) Substituting in using $t$: $\dfrac{1+t^2}{1-t^2} = \dfrac{4t}{1-t^2}(t) \Rightarrow 1+t^2 = 4t^2 \Rightarrow 3t^2 = 1 \Rightarrow t = \pm 1/\sqrt{3} \Rightarrow \tan(\tfrac{1}{2}x) = \pm 1/\sqrt{3}$,

$\tan(\tfrac{1}{2}x) = 1/\sqrt{3} \Rightarrow \tfrac{1}{2}x = 30°$, $\tan(\tfrac{1}{2}x) = -1/\sqrt{3} \Rightarrow \tfrac{1}{2}x = -30°$ $\Rightarrow x = 60°, -60°$

**4** LHS $\equiv \dfrac{1}{\cos x} + \dfrac{\sin x}{\cos x} \equiv \dfrac{1+\sin x}{\cos x}$

Multiplying top and bottom by $1-\sin x$: $\dfrac{1+\sin x}{\cos x} \times \dfrac{1-\sin x}{1-\sin x} \equiv \dfrac{1-\sin^2 x}{\cos x(1-\sin x)} \equiv \dfrac{\cos^2 x}{\cos x(1-\sin x)} \equiv \dfrac{\cos x}{(1-\sin x)} \equiv \text{RHS}$

## Differentiation

1a) $u = x+2, \quad v = e^{-2x} \Rightarrow \dfrac{du}{dx} = 1, \quad \dfrac{dv}{dx} = -2e^{-2x}$

Product rule $\Rightarrow \dfrac{dy}{dx} = e^{-2x} - 2(x+2)e^{-2x} = e^{-2x}(1-2x-4) = e^{-2x}(-2x-3)$

b) $\dfrac{dy}{dx} = 0 \Rightarrow e^{-2x}(-2x-3) = 0 \Rightarrow x = -3/2$

Hence $y = (-3/2+2)e^3 = \dfrac{1}{2}e^3 \Rightarrow$ co-ordinates are $C\left(-\dfrac{3}{2}, \dfrac{1}{2}e^3\right)$

c) $u = e^{-2x}, v = -2x-3 \Rightarrow \dfrac{du}{dx} = -2e^{-2x}, \dfrac{dv}{dx} = -2$

Product rule $\Rightarrow \dfrac{d^2y}{dx^2} = e^{-2x}(4x+6) - 2e^{-2x} = e^{-2x}(4x+4) = 4(x+1)e^{-2x}$

At C $\dfrac{d^2y}{dx^2} = -2e^3$ *Since* $\dfrac{d^2y}{dx^2} = -2e^3 < 0 \Rightarrow$ max S.P.

2a) $f'(x) = \cos x e^{\sin x}$

b) $u = \cos x, v = e^{\sin x} \Rightarrow \dfrac{du}{dx} = -\sin x, \dfrac{dv}{dx} = \cos x e^{\sin x}$

Product rule $\Rightarrow f''(x) = -\sin x e^{\sin x} + \cos^2 x e^{\sin x} = e^{\sin x}(\cos^2 x - \sin x)$

c) $f''(x) = 0 \Rightarrow e^{\sin x}(\cos^2 x - \sin x) = 0 \Rightarrow \cos^2 x - \sin x = 0 \Rightarrow 1 - \sin^2 x - \sin x = 0$

$\sin^2 x + \sin x - 1 = 0 \Rightarrow \sin x = \dfrac{-1 \pm \sqrt{1+4}}{2} = \dfrac{-1 \pm \sqrt{5}}{2}$

**3a)** Volume $V = 2x^2h \Rightarrow 1728 = 2x^2h \Rightarrow \dfrac{1728}{2x^2} = h$ ①

Surface area $A = y = xh + xh + 2xh + 2xh + 2x^2 \Rightarrow y = 2x^2 + 6xh$ ②

① into ② gives $y = 2x^2 + 6x\left(\dfrac{1728}{2x^2}\right) = 2x^2 + \left(\dfrac{6 \times 1728}{2x}\right) = 2x^2 + \dfrac{5184}{x}$

b) $\dfrac{dy}{dx} = 4x - \dfrac{5184}{x^2} = 0 \Rightarrow \dfrac{5184}{x^2} = 4x \Rightarrow x^3 = \dfrac{5184}{4} = 1296$

$x = \sqrt[3]{1296} = 10.9027\ldots = 10.9$ to three sig. figs.

$\dfrac{d^2y}{dx^2} = 4 + \dfrac{10368}{x^3}$ At $x = 10.9027\ldots \dfrac{d^2y}{dx^2} = 4 + \dfrac{10368}{1296} = 12 > 0$

$\Rightarrow$ minimum area at $x = 10.9$ cm

c) $A_{min} = 2(10.9027\ldots)^2 + \dfrac{5184}{10.9027\ldots} = 713.216\ldots = 713\text{ cm}^2$ to three sig. figs.

## Co-ordinate geometry with differentiation

**1a)** Gradient $= \dfrac{0-5}{0-2} = \dfrac{5}{2} \Rightarrow y = \dfrac{5}{2}x$

b) Equation $WX \Rightarrow y - 5 = -2(x - 2) \Rightarrow y = -2x + 9$

Grad $WX = -2 \Rightarrow$ grad $XY = \frac{1}{2}$ because lines are perpendicular

Equation $XY \Rightarrow y - 2 = \frac{1}{2}(x - 6) \Rightarrow y = \frac{1}{2}x - 1$

c) $WX = XY \Rightarrow -2x + 9 = \frac{1}{2}x - 1 \Rightarrow 10 = 2.5x \Rightarrow x = 4$

At $x = 4$, $y = -2(4) + 9 = 1$ So co-ordinates are $X(4, 1)$

d) Grad $OW$ = Grad $YZ = 5/2$

Equation $YZ \Rightarrow y - 2 = \dfrac{5}{2}(x - 6) \Rightarrow y = \dfrac{5}{2}x - 13$

When $y = 0$: $0 = \dfrac{5}{2}x - 13 \Rightarrow \dfrac{5}{2}x = 13 \Rightarrow x = \dfrac{26}{5} = 5.2$

Hence length of base $= 5.2 \times 20 = 104$ cm

**2a)** $x = 2t \quad y = \dfrac{6}{t} \Rightarrow \dfrac{dx}{dt} = 2 \quad \dfrac{dy}{dt} = \dfrac{-6}{t^2} \Rightarrow \dfrac{dy}{dx} = \dfrac{-3}{t^2}$

b) At $t = 3$, $x = 6$ and $y = 2$

At $t = 3$, grad of $T = \dfrac{-3}{3^2} = -\dfrac{1}{3} \Rightarrow$ grad of normal $= 3$

Equation of $N$: $y - 2 = 3(x - 6) \Rightarrow y = 3x - 16$

c) Curve = normal $\Rightarrow$ equate both equations $\Rightarrow \dfrac{6}{t} = 3(2t) - 16$

$\dfrac{6}{t} = 6t - 16 \Rightarrow 6 = 6t^2 - 16t \Rightarrow 6t^2 - 16t - 6 = 0 \Rightarrow 3t^2 - 8t - 3 = 0$

Factorising gives $(3t + 1)(t - 3) = 0$ So $t = 3, -1/3$

We know that $t = 3$ already so normal cuts curve again when $t = -1/3$

Hence $x = 2\left(\dfrac{-1}{3}\right) = -\dfrac{2}{3}$, $y = \dfrac{6}{t} = \dfrac{6}{\frac{-1}{3}} = -18$

The co-ordinates are $Q\left(-\dfrac{2}{3}, -18\right)$

## Probability and statistics

**1** True class boundaries are 39.5–49.5 etc, because weight is to nearest kg. So we have:

| Weight (kg) | Midpoint ($x$) | $f$ | $fx$ | $fx^2$ |
|---|---|---|---|---|
| 39.5–49.5 | 44.5 | 2 | 89 | 3960.5 |
| 49.5–54.5 | 52 | 10 | 520 | 27040 |
| 54.5–59.5 | 57 | 30 | 1710 | 97470 |
| 59.5–64.5 | 62 | 40 | 2480 | 153760 |
| 64.5–69.5 | 67 | 15 | 1005 | 67335 |
| 69.5–89.5 | 79.5 | 3 | 238.5 | 18960.75 |
| | | $\Sigma f = 100$ | $\Sigma fx = 6042.5$ | $\Sigma fx^2 = 368526.25$ |

a) Mean = 6042.5 ÷ 100 = 60.425 kg  SD = $\sqrt{(368526.25/100 - 60.425^2)}$ = 5.84 kg
These are only estimates because we haven't got the actual data values – the data have been grouped.

b)

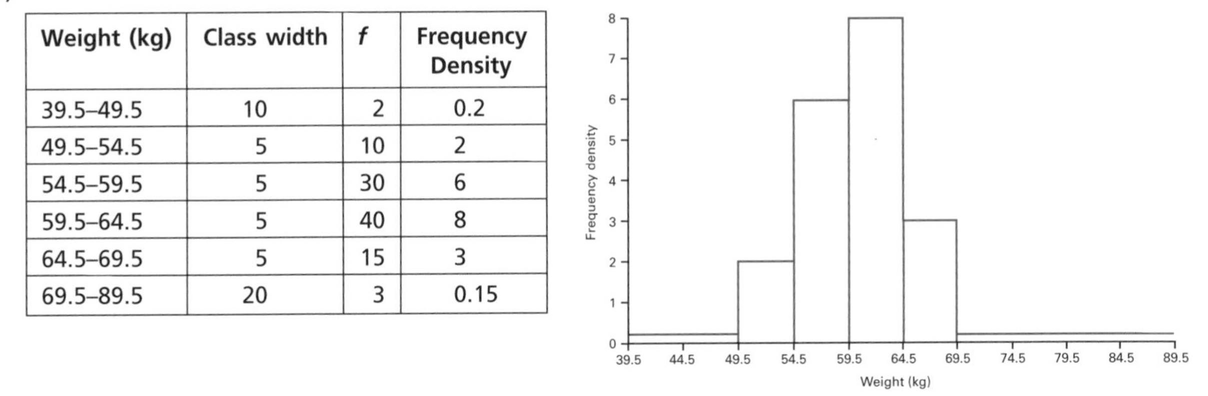

| Weight (kg) | Class width | $f$ | Frequency Density |
|---|---|---|---|
| 39.5–49.5 | 10 | 2 | 0.2 |
| 49.5–54.5 | 5 | 10 | 2 |
| 54.5–59.5 | 5 | 30 | 6 |
| 59.5–64.5 | 5 | 40 | 8 |
| 64.5–69.5 | 5 | 15 | 3 |
| 69.5–89.5 | 20 | 3 | 0.15 |



c) Need to multiply all data by $^{100}/_{110} = {}^{10}/_{11}$. So new mean = 60.425 kg × $^{10}/_{11}$ = 54.932 kg
New SD = 5.84 kg × $^{10}/_{11}$ = 5.31 kg

**2**a) $P$(positive) = 0.005 × 0.99 + 0.995 × 0.02 = 0.02485

b) This is $P$(disease|positive) = $P$(disease and positive)/$P$(positive)
= 0.005 × 0.99 ÷ 0.02485 = 0.199

c) The probability of her having the disease for which she has tested positive is only about 20 % – most of the positive tests are wrong ones – so it is not a very good test!

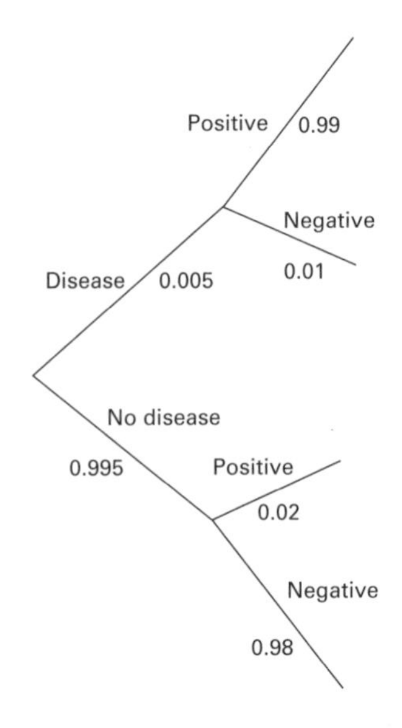

3a) $A \cap B$ means $A$ happens and $B$ happens. $A \cap B'$ means $A$ happens and $B$ doesn't happen. $P(A \cap B) + P(A \cap B') =$ prob [($A$ happens and $B$ happens) or ($A$ happens and $B$ doesn't happen)] $= P(A$ happens)

b) i) We know that all girls play netball or hockey or both. So girls either play just netball, netball and hockey or just hockey. Just netball is $N \cap H'$. Just hockey is $N' \cap H$. Hockey and netball is the same as $N \cap H$
So $P(N \cap H') + P(N' \cap H) + P(N \cap H) = 1$

ii) We know $P(N|H) = \frac{1}{4}$ $P(N) = \frac{11}{20}$ We want $P(\text{just hockey}) = P(N' \cap H)$
Try using $P(N \cap H') + P(N' \cap H) + P(N \cap H) = 1 \Rightarrow P(N' \cap H) = 1 - [P(N \cap H') + P(N \cap H)]$
But $P(N) = P(N \cap H) + P(N \cap H')$, from what was shown in a)
So $P(N' \cap H) = 1 - P(N) = \frac{9}{20}$

iii) We know from a) $P(H) = P(N \cap H) + P(N' \cap H)$
$P(N' \cap H) = \frac{9}{20}$ $P(H) = h$.
Have not yet used $P(N|H) = P(N \cap H) \div P(H) \Rightarrow \frac{1}{4} = P(N \cap H) \div h \Rightarrow P(N \cap H) = \frac{1}{4}h$
So, substituting in: $h = \frac{1}{4}h + \frac{9}{20} \Rightarrow \frac{3}{4}h = \frac{9}{20} \Rightarrow h = \frac{3}{5}$

iv) $P(\text{not hockey}|\text{netball}) = P(\text{not hockey and netball}) \div P(\text{netball})$
So we need $P(H' \cap N) = P(N) - P(H \cap N)$
But $P(N) = \frac{11}{20}$; $P(H \cap N) = \frac{1}{4}h = \frac{3}{20}$
So $P(H' \cap N) = \frac{11}{20} - \frac{3}{20} = \frac{2}{5}$
So $P(\text{not hockey}|\text{netball}) = \frac{2}{5} \div \frac{11}{20} = \frac{8}{11}$

## Integration

**1** a) $-\ln|2-x| + c$

b) $y = \int 4 - \frac{1}{x^3} + \frac{1}{2-x}\,dx = 4x + \frac{1}{2x^2} - \ln(2-x) + c$

when $x = 1, y = \frac{4}{3} \Rightarrow \frac{5}{2} = 4 + \frac{1}{2} - 0 + c \Rightarrow c = -2$

Hence $y = 4x + \frac{1}{2x^2} - \ln(2-x) - 2$.

**2** a) $8x + 7 = (Ax + B)(4 - x) + C(2x^2 + 7)$

When $x = 4$, $39 = 39C \Rightarrow C = 1$

When $x = 0$, $7 = 4B + 7C \Rightarrow 7 - 7 = 4B \Rightarrow B = 0$

When $x = 1$, $15 = 3A + 9C \Rightarrow 6 = 3A \Rightarrow A = 2$

Hence $f(x) = \frac{2x}{2x^2 + 7} + \frac{1}{4 - x}$

b) $\int \frac{2x}{2x^2+7} + \frac{1}{4-x}\,dx = \left[\frac{1}{2}\ln(2x^2+7) - \ln(4-x)\right]_1^3$

$\left(\frac{1}{2}\ln 25 - \ln 1\right) - \left(\frac{1}{2}\ln 9 - \ln 3\right) = \ln 5 - 0 - \ln 3 + \ln 3 = \ln 5$

Hence $D = 5$

**3** a) $u = x$ $\frac{dv}{dx} = \cos 3x \Rightarrow \frac{du}{dx} = 1$ $v = \frac{1}{3}\sin 3x$

$\int x\cos x\,dx = \frac{x}{3}\sin 3x - \int \frac{1}{3}\sin 3x\,dx = \frac{x}{3}\sin 3x + \frac{1}{9}\cos 3x + c$

b) $x = 3\sin u \Rightarrow \frac{dx}{du} = 3\cos u \Rightarrow dx = 3\cos u du$

Changing limits $\Rightarrow$ when $x = 0$, $0 = 3\sin u \Rightarrow u = 0$

When $x = \frac{3}{2}$, $\frac{3}{2} = 3\sin u \Rightarrow \frac{1}{2} = \sin u \Rightarrow u = \sin^{-1}\left(\frac{1}{2}\right) = \frac{\pi}{6}$

$$\int_0^{3/2} \sqrt{(9-x^2)}\, dx = \int_0^{\pi/6} \sqrt{(9-9\sin^2 u)}\ 3\cos u du = \int_0^{\pi/6} \sqrt{9(1-\sin^2 u)}\ 3\cos u du$$

$$= \int_0^{\pi/6} \sqrt{9\cos^2 u}\, 3\cos u du = \int_0^{\pi/6} 3\cos u\ 3\cos u du = 9\int_0^{\pi/6} \cos^2 u du$$

Hence $k = 9$, $a = \frac{\pi}{6}$

c) From trig, $\cos 2u = 2\cos^2 u - 1 \Rightarrow \cos^2 u = \frac{\cos 2u + 1}{2}$

The integral in b) becomes $9\int_0^{\pi/6} \frac{\cos 2u + 1}{2} du = \frac{9}{2}\left[\frac{\sin 2u}{2} + u\right]_0^{\pi/6}$

$= \frac{9}{2}\left(\left(\frac{\sqrt{3}}{4} + \frac{\pi}{6}\right) - (0)\right) = \frac{9}{8}\sqrt{3} + \frac{3}{4}\pi$ Hence $e = \frac{9}{8}$ and $f = \frac{3}{4}$

**4** a) $u = w\ \frac{dv}{dx} = e^{-w} \Rightarrow \frac{du}{dx} = 1 \quad v = -e^{-w}$

$\int we^{-w} dw = -we^{-w} + \int e^{-w} dw = -we^{-w} - e^{-w} + c$

b) $x = e^w \Rightarrow \frac{dx}{dw} = e^w \Rightarrow dx = e^w dw$ Note that $x^2 = e^w e^w$

$\int \frac{3 - \ln x}{x^2} dx = \int \frac{3 - \ln e^w}{e^w e^w} e^w dw = \int \frac{3 - w}{e^w} dw$, as required

c) $\int \frac{3 - w}{e^w} dw = \int \frac{3}{e^w} - \frac{w}{e^w} dw = \int 3e^{-w} - we^{-w} dw$

Using part a) the integral becomes: $= -3e^{-w} + we^{-w} + e^{-w} + c = we^{-w} - 2e^{-w} + c$

$= \frac{w}{e^w} - \frac{2}{e^w} + c$ Since $x = e^w$ substituting back gives $= \frac{\ln x}{x} - \frac{2}{x} + c$

## Applications of integration

**1** a) $W(1, 0)$, $X(5, 0)$, $Y(2, -3/4)$, $Z(4, -\frac{1}{2})$

b) $\frac{(x-1)(x-5)}{(x+2)} = (Ax + B) + \frac{C}{(x+2)} \Rightarrow (x-1)(x-5) = (Ax+B)(x+2) + C$

When $x = -2$, $(-3)(-7) = C \Rightarrow C = 21$

When $x = 0$, $(-1)(-5) = 2B + C \Rightarrow 2B + C = 5 \Rightarrow 2B = -16 \Rightarrow B = -8$

When $x = 1$, $(0)(-4) = 3A + 3B + C \Rightarrow 3A - 24 + 21 = 0 \Rightarrow A = 1$

Hence $y = x - 8 + \frac{21}{(x+2)}$

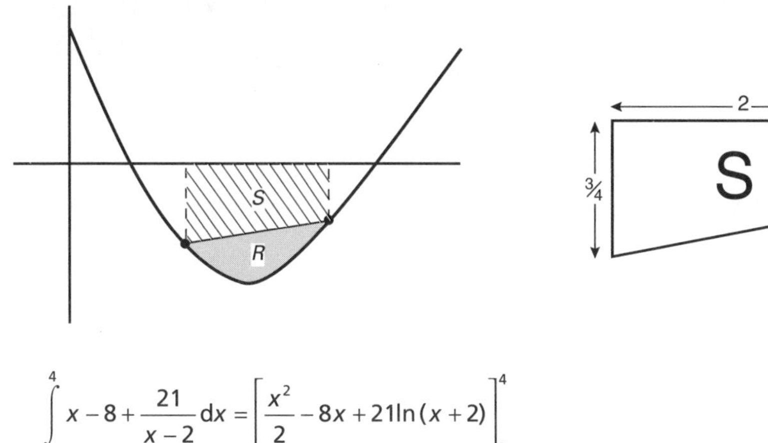


$$\int_2^4 x - 8 + \frac{21}{x-2}\,\mathrm{d}x = \left[\frac{x^2}{2} - 8x + 21\ln(x+2)\right]_2^4$$

$$= (8 - 32 + 21\ln 6) - (2 - 16 + 21\ln 4) = -24 + 21\ln 6 + 14 - 21\ln 4$$

$$= -10 + 21\ln 6 - 21\ln 4 = -10 + 21\ln\left(\frac{3}{2}\right)$$

c) Since we are integrating below the $x$-axis then area $(R + S) = 10 - 21\ln\left(\frac{3}{2}\right)$

Region $S$ is a trapezium $\Rightarrow$ Area $(S) = \frac{1}{2}\left(\frac{3}{4} + \frac{1}{2}\right)2 = \frac{5}{4}$

Area $(R)$ = area $(R + S)$ – area $(S)$ $= 10 - 21\ln\left(\frac{3}{2}\right) - \frac{5}{4} = \frac{35}{4} - 21\ln\left(\frac{3}{2}\right)$

**2** a) $u = x^2 \quad \frac{\mathrm{d}v}{\mathrm{d}x} = \mathrm{e}^{-3x} \Rightarrow \frac{\mathrm{d}u}{\mathrm{d}x} = 2x \quad v = -\frac{1}{3}\mathrm{e}^{-3x}$

$\int x^2\mathrm{e}^{-3x}\mathrm{d}x = -\frac{1}{3}x^2\mathrm{e}^{-3x} + \frac{2}{3}\int x\mathrm{e}^{-3x}\mathrm{d}x \Rightarrow$ Apply by parts on $x\mathrm{e}^{-3x}$

$u = x \quad \frac{\mathrm{d}v}{\mathrm{d}x} = \mathrm{e}^{-3x} \Rightarrow \frac{\mathrm{d}u}{\mathrm{d}x} = 1 \quad v = -\frac{1}{3}\mathrm{e}^{-3x}$

$\int x\mathrm{e}^{-3x}\mathrm{d}x = -\frac{x}{3}\mathrm{e}^{-3x} - \frac{1}{3}\int \mathrm{e}^{-3x}\mathrm{d}x = -\frac{1}{3}x\mathrm{e}^{-3x} - \frac{1}{9}\mathrm{e}^{-3x}.$

Hence $\int x^2\mathrm{e}^{-3x}\mathrm{d}x = -\frac{1}{3}x^2\mathrm{e}^{-3x} + \frac{2}{3}\left(-\frac{x}{3}\mathrm{e}^{-3x} - \frac{1}{9}\mathrm{e}^{-3x}\right) + c$

$$= -\frac{1}{3}x^2\mathrm{e}^{-3x} - \frac{2}{9}x\mathrm{e}^{-3x} - \frac{2}{27}\mathrm{e}^{-3x} + c$$

$$= -\frac{1}{27}\mathrm{e}^{-3x}(9x^2 + 6x + 2) + c \Rightarrow f(x) = 9x^2 + 6x + 2$$

b) $\frac{\mathrm{d}y}{\mathrm{d}x} = x^2\mathrm{e}^{y-3x} \Rightarrow \frac{\mathrm{d}y}{\mathrm{d}x} = x^2\mathrm{e}^{y}\mathrm{e}^{-3x} \Rightarrow \int \frac{1}{\mathrm{e}^y}\mathrm{d}y = \int x^2\mathrm{e}^{-3x}\mathrm{d}x \Rightarrow$

$\int \mathrm{e}^{-y}\mathrm{d}y = \int x^2\mathrm{e}^{-3x}\mathrm{d}x \Rightarrow -\mathrm{e}^{-y} = -\frac{1}{27}\mathrm{e}^{-3x}(9x^2 + 6x + 2) + c$

When $x = 0, y = 0 \Rightarrow -1 = -\frac{1}{27}(2) + c \Rightarrow c = -\frac{25}{27}$

Hence $-\mathrm{e}^{-y} = -\frac{1}{27}\mathrm{e}^{-3x}(9x^2 + 6x + 2) - \frac{25}{27}$

## Numerical methods

**1** a) $x = A + \frac{B}{x+1} \Rightarrow x(x+1) = A(x+1) + B \Rightarrow x^2 + x = Ax + A + B \Rightarrow x^2 + x - Ax - A - B = 0$

$\Rightarrow x^2 + (1-A)x - (A+B) = 0$

So $1 - A = 0$, giving $A = 1$ and $A + B = 2$, giving $B = 1$

b) $x_{n+1} = 1 + \frac{1}{x_n + 1}$ $x_1 = 1 \Rightarrow x_2 = 1 + \frac{1}{1+1} = 1.5 \Rightarrow x_3 = 1 + \frac{1}{1.5+1} = 1.4 \Rightarrow x_4 = 1 + \frac{1}{1.4+1} = 1.41667$

$\Rightarrow x_5 = 1 + \frac{1}{1.41667+1} = 1.41379 \Rightarrow x_6 = 1 + \frac{1}{1.41379+1} = 1.41429 \Rightarrow x_7 = 1 + \frac{1}{1.41429+1} = 1.41420$

$\Rightarrow x_8 = 1 + \frac{1}{1.41420+1} = 1.41422 \Rightarrow$ answer is 1.4142 (4 D.P.)

**2** a) i)

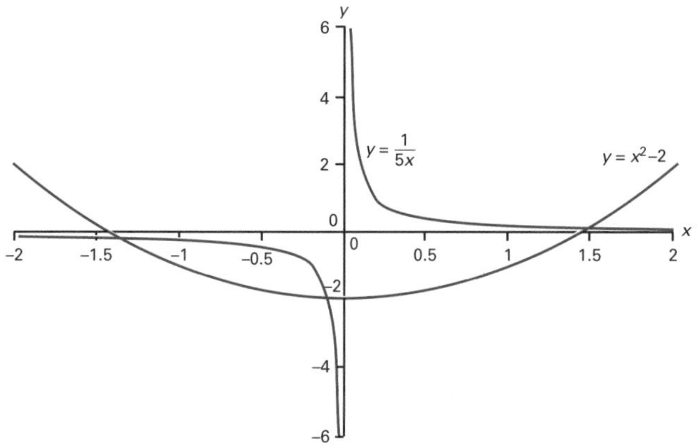

ii) $5x(x^2 - 2) = 1 \Rightarrow x^2 - 2 = \frac{1}{5x}$ From the graph, there are three solutions

b) $5x(x^2 - 2) = 1 \Rightarrow 5x^3 - 10x - 1 = 0$ At $x = 0$, $5x^3 - 10x - 1 = -1$ At $x = -1$, $5x^3 - 10x - 1 = 4$

Sign change $\Rightarrow$ root

$x = -0.6$: $5x^3 - 10x - 1 = 3.92$ $x = -0.2$: $5x^3 - 10x - 1 = 0.96$ $x = -0.1$: $5x^3 - 10x - 1 = -0.005$

So root is in between $-0.2$ and $-0.1$ = root is $-0.2$ to nearest 0.2

**3** a)

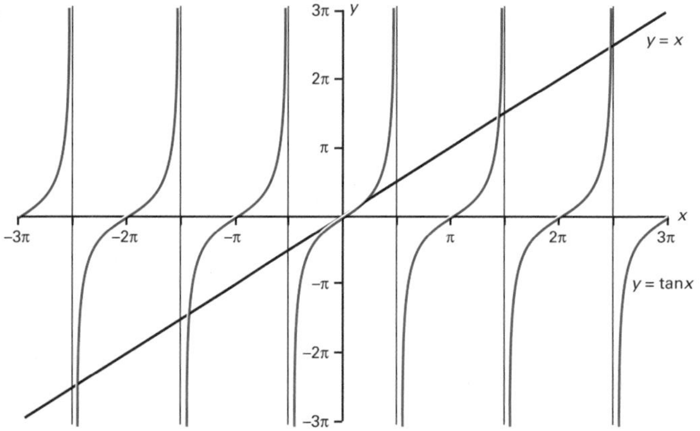

$y = x$ crosses $y = \tan x$ once in every cycle of $\tan x$, and there is not an upper or lower limit on the values $\tan x$ takes – so it continues to cross it once every $\pi$ radians – so it crosses infinitely many times.

b) We want roots of $\tan x - x = 0$. $\pi$ is about 3.14, and $\frac{3\pi}{2}$ is about 4.71, so we're looking for roots between 3.14 and 4.71

$x = 3.2$: $\tan x - x = -3.14$ $x = 4.6$: $\tan x - x = 4.26$ $x = 3.8$: $\tan x - x = -3.03$

$x = 4.2$: $\tan x - x = -2.42$ $x = 4.4$: $\tan x - x = -1.30$ $x = 4.5$: $\tan x - x = 0.14$

Root is between 4.4 and 4.5 – so it is 4.4 to nearest 0.2

c) i) $x_1 = 4.4$ $x_2 = \tan^{-1}4.4 = 1.34732$ $x_3 = \tan^{-1}1.34732 = 0.932297$

ii) The values are moving away – diverging – from the root we want, so the formula is not suitable.